CHEMISTRY: A MODERN INTRODUCTION
FOR SIXTH FORMS AND COLLEGES

PART 3

Physical Chemistry

CHEMISTRY: A MODERN INTRODUCTION FOR SIXTH FORMS AND COLLEGES

PART 3
Physical Chemistry

T. F. Chadwick, B.Sc., F.R.I.C., Dip. Chem. Eng.

Swinburne Institute of Technology,
Hawthorn, Victoria, Australia
Formerly Senior Lecturer in Chemistry,
Blackburn College of Technology and Design

London
GEORGE ALLEN & UNWIN LTD
RUSKIN HOUSE   MUSEUM STREET

First published in 1972

This book is copyright under the Berne Convention. All rights are reserved. Apart from any fair dealing for the purpose of private study, research, criticism or review, as permitted under the Copyright Act, 1956, no part of this publication may be reproduced, stored in a retrieval system, or transmitted, in any form or by any means, electronic, electrical, chemical, mechanical, optical, photocopying, recording or otherwise, without the prior permission of the copyright owner. Enquiries should be addressed to the publishers.

© *George Allen & Unwin Ltd 1972*

ISBN 0 04 541003 8

*Printed in Great Britain
in 10 on 11 Point Times Roman
by Alden & Mowbray Ltd
at the Alden Press, Oxford*

PREFACE

The object of this book is to present, in a concise manner, an introduction to the study of physical chemistry. The subject matter is intended to meet the requirements of the 'A' and 'S' levels of the G.C.E. examination and the Ordinary National Certificate in Sciences.

The book opens with a consideration of the scope of physical chemistry, the basic quantitative laws and the mole concept. The role of energetics in chemistry at this level is covered in some detail in Chapter 4, while the discussion of redox potentials in Chapter 6 forms a link with some of the material presented in Part 1 of this series, *General and Inorganic Chemistry*. Most of the units appearing in the book agree with SI requirements, and an explanation of the application of SI units to chemistry appears in Chapter 4.

An ability to handle the mathematics of the subject is essential for a real understanding of physical chemistry and, in the author's experience, it is this aspect that presents many students with difficulties. In consequence, eighty numerical examples are worked through at appropriate points in the text, while further exercises have been added at the ends of most chapters. The solutions to these calculations are given at the end of the book. (All calculations have been checked at least twice, but, if any error still remains, the author would be glad to have it brought to his notice.)

A short reading list and sources of data are indicated in the appendix, and reference could profitably be made to articles appearing in journals and periodicals such as *Education in Chemistry*, *The Journal of Chemical Education*, *The School Science Review*, *Chemistry in Britain*, *Endeavour* and others.

Finally, may I express my appreciation to the publishers and many others who have given their valuable service and assistance in connection with the preparation of this series.

Hawthorn, Victoria, 1971 T.F.C.

CONTENTS

Preface	*page* 7
1. Introduction: scope and basic laws of physical chemistry	11
2. Gases	20
3. Liquids and solutions	54
4. Energy changes in chemistry	102
5. Chemical reactions and equilibrium	146
6. Electrochemistry and ionic equilibrium	177
7. The colloidal state	225
8. Radioactivity	232

APPENDIX:

Atomic weights of the elements	243
Lattice energies	244
Electron affinities	244
Hydration energies	244
Ionic radii	244
Bond energies	244
Short reading list	245
Answers to exercises	246
Tables of common logarithms	248
Index	252

1. Introduction

The Scope of Physical Chemistry

Inorganic and organic chemistry are concerned with the changes that take place in the chemical constitution of materials following a chemical reaction. Physical chemistry attempts to measure the influence of factors such as heat, light, pressure, concentration, electricity, etc., on both the reactants and the reaction itself, and to deduce from these measurements the fundamental laws governing chemical change. Thus, physical chemistry is concerned with:

1. collecting data regarding the physical properties of materials, and the design of apparatus and techniques suitable for the accurate measurement of such data;
2. measuring energy exchanges relating to chemical reactions, determining the speed of chemical reactions and the extent to which they take place, and studying the effect of changes in the various factors listed above on the reaction;
3. correlating the information obtained, assessing its significance and systematizing it in the form of laws;
4. developing a theoretical explanation of these laws, and making deductions from them, leading to an advancement of the subject in accordance with the scientific method (see Part 1).

Physical chemistry uses the methods and laws of physics and mathematics to investigate chemical systems on a macro scale and to interpret the observations theoretically in terms of molecular behaviour.

The Basic Laws of Physical Chemistry

The results of the application of quantitative methods in the earlier development of chemistry are summarized by the following laws.

The law of conservation of mass (Lavoisier, 1774)   *Matter can neither be created nor destroyed.*

The law of constant composition (Proust, 1799)   *Pure compounds always contain the same elements in the same proportions by weight.*

The law of multiple proportions (Dalton, 1803) *When two elements combine to form more than one compound, the weights of one element which combine with a fixed weight of the other are in simple whole number ratio.*

The law of reciprocal proportions (Richter, 1792) *Elements combine with, or displace one another in relative quantities which are simply related to their equivalents.*

The law of conservation of energy *Energy may not be created or destroyed, although it may change its form.* (Both this law and the law of conservation of mass apply only to chemical reactions. In a nuclear reaction, mass is in fact converted into energy.)

The Atomic and Molecular Theories

The laws of chemical combination (constant composition, multiple proportions and reciprocal proportions) were explained by *Dalton's atomic theory* (1807). That matter could be regarded as a collection of small indivisible particles called atoms was an idea of very long standing, but Dalton developed the concept into an important theory by suggesting that:

(a) atoms have fixed weights, all atoms of the same element having the same weight while atoms of different elements have different weights;

(b) compounds are formed by the combination of atoms in a fixed ratio of whole (and usually small) numbers.

Avogadro, in 1811, drew the distinction between atoms and molecules: an atom is the smallest particle that can take part in a chemical reaction, and a molecule is the smallest particle that can have a free existence. The atomicity of a molecule is equal to the number of atoms it contains.

The concept of a whole number relationship between the atoms in a compound, or the molecules taking part in a reaction, is termed *stoichiometry*. There are many compounds in which a whole number relationship between atoms is not found, for example $FeO_{1.08}$. These compounds are termed *non-stoichiometric*. In addition, many compounds, especially organic compounds, are known to contain many hundreds of atoms in each molecule, so that the generalization that atoms combine in the ratio of small whole numbers is somewhat outdated. As with all theories, Dalton's atomic theory has to be modified and extended to agree with new developments, although it still remains a simple, useful and important concept.

The molecular theory was introduced by Avogadro to explain the following discovery:

Gay Lussac's law of combining volumes (1808) *The volumes of gases which react, and the volumes of the products, if gaseous (when measured under identical conditions) are in the ratio of simple whole numbers.*

For example, in the reaction between chlorine and hydrogen it was found that:

   1 volume of chlorine + 1 volume of hydrogen
   → 2 volumes of hydrogen chloride

This was explained as follows.

Avogadro's hypothesis (1811) *Equal volumes of gases (measured under identical conditions) contain equal numbers of molecules.*

Thus, in the example given above,

   n molecules of chlorine + n molecules of hydrogen
   → $2n$ molecules of hydrogen chloride

Cancelling throughout by n,

   1 molecule of chlorine + 1 molecule of hydrogen
   → 2 molecules of hydrogen chloride

Since the hydrogen chloride molecule must contain at least one hydrogen and one chlorine atom, then both hydrogen and chlorine molecules must each contain at least two atoms per molecule.

Evidence such as the following facts indicates that the hydrogen molecule is diatomic:

(a) one volume of hydrogen has never been found to give rise to more than two volumes of gaseous product in any reaction;
(b) the acid produced on hydrating hydrogen chloride is monobasic;
(c) the ratio of the molar heat capacities of hydrogen (see p. 106) is 1·4.

Atomic and Molecular Weights*

The atomic weight of an element is a relative quantity, being the ratio of the weight of one atom of an element to the weight of one atom of a reference element. Modern methods of determining atomic weights depend on the use of a mass spectrograph (p. 238), and the present definition of atomic weight is made in accordance with this technique:

$$\text{atomic weight} = \frac{\text{weight of one atom of element}}{\text{weight of one atom of carbon-12 isotope}} \times 12$$

* These terms are currently being replaced by relative atomic mass and relative molecular mass (or formula weight) respectively.

The multiplying factor 12 is introduced to avoid having fractional weights for atoms lighter than carbon.

The molecular weight of a substance may be defined either in a similar manner to the atomic weight,

$$\text{molecular weight} = \frac{\text{weight of 1 molecule of the substance}}{\text{weight of 1 atom of carbon-12 isotope}} \times 12$$

or more simply as the sum of the atomic weights of all the atoms shown in the molecular formula of the substance. Thus, the molecular weight of toluene, $C_6H_5CH_3$, is

$$6 \times 12 \cdot 011 + 5 \times 1 \cdot 008 + 1 \times 12 \cdot 011 + 3 \times 1 \cdot 008 = 92 \cdot 141$$

Many substances do not form molecular entities. Silica, for example, consists of a giant network of silicon and oxygen atoms bonded together in a regular manner, while sodium fluoride does not contain discrete molecules of NaF, but is made up of a regular assembly of sodium and fluoride ions. In such cases, it is incorrect to use the term molecular weight, since no molecules are present. Instead, the term *mole* has been introduced in order to express quantity without reference to actual molecules. *One mole is the number of atoms in exactly 12 g of the carbon-12 isotope.* This number, $6 \cdot 02 \times 10^{23}$, is known as *Avogadro's number*, and it is given the symbol N. (A millimole is 10^{-3} mole, and a micromole is 10^{-6} mole.) Alternatively, *one mole is that quantity of material which contains N chemical entities or units.* In calculating these quantities, the exact nature of the chemical entity must be clear, as shown in the following examples.

Example 1 Calculate the weight of (a) 0·4 mole of chlorine gas, (b) 0·2 mole of chloride ion.

Solution (a) The chemical units present in chlorine gas are Cl_2 molecules. 1 mole of these molecules weighs

$$2 \times 35 \cdot 5 = 71 \text{ g}$$

(using the approximate atomic weights listed at the end of the book). Therefore, 0·4 mole weighs

$$0 \cdot 4 \times 71 = 28 \cdot 4 \text{ g}$$

(b) The chemical units in this case are Cl^- ions. 1 mole Cl^- ions weighs

$$35 \cdot 5 \text{ g}$$

and therefore 0·2 mole weighs

$$0 \cdot 2 \times 35 \cdot 5 = 7 \cdot 1 \text{ g}$$

Example 2 What is the weight of 12.04×10^{20} atoms of neon?

Solution 1 mole (6.02×10^{23} atoms) of neon weighs 20·2 g. Therefore 12.04×10^{20} atoms weigh

$$\frac{20.2 \times 12.04 \times 10^{20}}{6.02 \times 10^{23}} = 4.04 \times 10^{-2} \text{ g}$$
$$= 0.0404 \text{ g}$$

Empirical, Molecular and Structural Formulae

The *empirical formula* of a compound shows the simplest mole ratio in which the elements are present.

Example 3 A pure compound is shown by analysis to contain 40% carbon, 53·3% oxygen and 6·7% hydrogen. Deduce the empirical formula for the compound.

Solution Let the formula be $C_xH_yO_z$. In 100 g of this compound,

$$\text{the number of moles of carbon, } x = \frac{40}{12} = 3.33$$

$$\text{the number of moles of hydrogen, } y = \frac{6.7}{1} = 6.7$$

$$\text{the number of moles of oxygen, } z = \frac{53.3}{16} = 3.33$$

Thus $x:y:z = 1:2:1$, and the empirical formula is CH_2O.

The *molecular formula* applies only to substances which are present as discrete molecules. Provided the weight of one mole of compound is known, then the total number of atoms in each molecule can be deduced. The formula showing this is the molecular formula. The ratio of the numbers of atoms in the molecular formula must be the same as that in the empirical formula, and thus the molecular formula is some multiple of the empirical formula.

Example 4 If 1 mole of the compound analysed in Example 3 weighs 60 g, deduce its molecular formula.

Solution 1 mole of a compound of empirical formula CH_2O would weigh $12+2+16 = 30$ g. Hence, the empirical formula corresponds to $30/60 = 0.5$ mole, and the molecular formula is thus $C_2H_4O_2$.

The *structural formula* shows the order in which the atoms are linked together in a molecule, and a knowledge of molecular structure is vitally important in the study of chemistry (see Parts 1 and 2).

In the example just given, the structural formula could be either:

$$\underset{\text{acetic acid}}{\text{H}-\overset{\overset{\text{H}}{|}}{\underset{\underset{\text{H}}{|}}{\text{C}}}-\overset{\nearrow^{\text{O}}}{\underset{\searrow_{\text{O}-\text{H}}}{\text{C}}}} \quad \text{or} \quad \underset{\text{methyl formate}}{\text{H}-\text{C}\overset{\nearrow^{\text{O}-\text{H}}}{\underset{\searrow_{\text{O}-\overset{\overset{\text{H}}{|}}{\underset{\underset{\text{H}}{|}}{\text{C}}}-\text{H}}}{}}}$$

Example 5 On heating 3·52 g of barium chloride crystals, 3·01 g of the anhydrous salt was produced. Calculate the degree of hydration of the crystals.

Solution

$$\text{Number of moles of BaCl}_2 = \frac{3 \cdot 01}{208 \cdot 3} = 0 \cdot 0144$$

$$\text{Number of moles of H}_2\text{O} = \frac{0 \cdot 51}{18} = 0 \cdot 0283$$

Hence,

$$\text{BaCl}_2 : \text{H}_2\text{O} = 0 \cdot 0144 : 0 \cdot 0283 = 1 : 2$$

Therefore the crystals contain two moles of water of crystallization per mole of barum chloride.

Molar Solutions

A molar solution is defined as containing one mole of a specified reagent per litre of solution.

A solution containing 1 mole per litre is designated M, a solution containing $\frac{1}{10}$ mole per litre is designated M/10, and in the general case, a solution containing f moles per litre is designated f.M (f is known as the *factor* of the solution).

From this definition,

$$1000 \text{ cm}^3 \text{ of an } f.\text{M solution contains } f \text{ moles}$$

and by proportion,

$$V \text{ cm}^3 \text{ of an } f.\text{M solution contains } \frac{V}{1000} f \text{ moles}$$

Example 6 Calculate the number of moles of (a) calcium bromide in 100 cm^3 of 0·3M solution, (b) bromide ion in the same solution.

Solution (a) Number of moles of calcium bromide is

$$\frac{V}{1000} \times f = \frac{100}{1000} \times 0\cdot 3 = 0\cdot 03 \text{ mole}$$

(b) Since each mole of calcium bromide, $CaBr_2$, contains 2 moles of bromide ion, the number of moles of bromide ion in this solution is

$$0\cdot 03 \times 2 = 0\cdot 06 \text{ mole}$$

Example 7 It was found that 25 cm^3 of M/10 phosphorous acid (H_3PO_3) neutralized exactly 40 cm^3 of a solution containing 5 g of sodium hydroxide per litre. Deduce the equation for the reaction between sodium hydroxide and phosphorous acid.

Solution

$$\text{Number of moles of } H_3PO_3 = \frac{25}{1000} \times \frac{1}{10} = 2\cdot 5 \times 10^{-3}$$

$$\text{Number of moles of NaOH} = \frac{40}{1000} \times \frac{5}{40} = 5 \times 10^{-3}$$

Therefore, the mole ratio of the reactants is

$$NaOH:H_3PO_3 = 5 \times 10^{-3} : 2\cdot 5 \times 10^{-3} = 2:1$$

and the equation for the reaction is

$$2NaOH + H_3PO_3 = Na_2HPO_3 + 2H_2O$$

EXERCISES

(Use the approximate atomic weights given in the table on page 243, at the end of the book.)

1. Calculate the weights of one mole of each of the following:
   (a) oxalic acid ($H_2C_2O_4$)
   (b) oxalic acid crystals, which contain two moles of water of crystallization
   (c) phosphorus molecules (P_4)

2. What weight corresponds to the following?
   (a) 0·3 mole hydrogen ion (H^+)
   (b) 0·3 mole hydride ion (H^-)
   (c) 0·6 mole hydrogen gas
   (d) one millimole hydrated copper (II) sulphate ($CuSO_4.5H_2O$)
   (e) $12\cdot 04 \times 10^{20}$ atoms of calcium

3. How many moles do the following quantities represent?

   (a) 2 g ice
   (b) 3.01×10^{12} alpha particles
   (c) 16 g oxygen gas
   (d) 16 g oxide ion
   (e) 48 g ozone
   (f) 20 cm^3 benzene (density 0.87 g cm^{-3})

4. Deduce the empirical formula of a gaseous hydrocarbon which contains 92.4% carbon. If one mole of this hydrocarbon weighs 26.0 g, what is its molecular formula?

5. A liquid hydrocarbon has the same empirical formula as the compound specified in Question 4. What is the molecular formula for this substance if three millimoles weigh 0.234 g?

6. When 1 millimole of a chloride of phosphorus was completely decomposed by water, 0.430 g of silver chloride was precipitated on the addition of excess silver nitrate solution (in the presence of dilute nitric acid) to this solution. What is the formula for the chloride of phosphorus used in the experiment?

7. What weight of sulphate ion is present in 100 cm^3 of the following solutions?

   (a) 0.2M sulphuric acid
   (b) 0.5M iron (II) ammonium sulphate
   (c) 0.01M potassium aluminium sulphate
   (d) 0.001M tetrammine copper (II) sulphate, $[Cu(NH_3)_4]SO_4$

8. Calculate the molarity of the following:

   (a) a solution containing 4.25 g sodium nitrate per 100 cm^3
   (b) a solution whose concentration is 8.3 g potassium iodide per 500 cm^3
   (c) a solution in which 1 cm^3 contains 0.01 g sodium benzoate (C_6H_5COONa)

9. A solution of sodium hydroxide contains 8 g NaOH per litre. If 2 cm^3 of this solution is mixed with 8 cm^3 of 1.2M sodium hydroxide, what is the molarity of the resulting solution? (Assume no change in total volume takes place on mixing.)

10. Deduce the equation for the reaction between a metal M and hydrochloric acid if 1 millimole of metal reacts with exactly 16.7 cm^3 0.12M acid.

11. What is the molarity of the sodium chloride solution produced by the neutralization of 10 cm^3 of 0.084M sodium hydroxide by 0.126M hydrochloric acid?

12. If 20 cm³ of 0·08M hydrochloric acid were needed to discharge the colour produced by adding two drops of phenolphthalein to 16 cm³ of 0·1M sodium carbonate solution, deduce the equation for the neutralization of sodium carbonate under these conditions.

2. Gases

States of Matter

Matter can be classified into three main divisions, namely, gases, liquids and solids.

Gases have neither shape nor size but fill completely any space into which they are introduced. All gases mix completely and a gas exerts pressure equally in all directions. The volume of a gas changes significantly corresponding to changes in temperature and pressure.

Liquids have a definite volume but no shape. They lack rigidity and elasticity but possess a boundary surface which confers many characteristic properties on them. Therefore, when placed in contact they may either diffuse completely (fully miscible) or mix to a limited or negligible extent. The change in the volume of a liquid brought about by an alteration in temperature is much less marked than with gases, while the effect of pressure is almost negligible.

Solids have both size and shape. They resist distortion (i.e. they are rigid) or recover from slight deformations (i.e. they are elastic) when the deforming force is removed. The volume of a solid does not change to any great extent following changes in temperature and pressure. These properties may be explained in terms of the closer packing and increasing intermolecular forces as the physical state changes from gaseous through liquid to solid.

The Gas Laws

The volume of a gas changes considerably as a result of variations in temperature and pressure. The effects of these changes are summarized by the laws of Boyle and Charles.

Boyle's law *The volume of a given mass of gas is inversely proportional to the pressure, at a constant temperature.*

Taking V to represent the volume and P to represent the total pressure on the gas, the law is expressed by the relation

$$V \propto \frac{1}{P} \quad \text{or} \quad PV = \text{constant}$$

The relationship is expressed graphically in Figure 1. Each curve is an isothermal, that is the temperature is the same for all the data lying on a single curve.

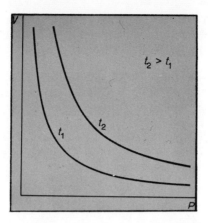

Fig. 1. Pressure–volume curve for a gas at a constant temperature

Alternatively, the product PV may be plotted against P. According to Boyle's law, this product should have a constant value at all pressures, as indicated by the horizontal line in Figure 2. In fact, the actual behaviour of any gas deviates considerably from this requirement; consequently, the evaluation of a constant value for the product PV at all pressures is an attribute of an *ideal* or *perfect gas*.

Boyle's law was based on a restricted number of observations of limited accuracy. Later work by Regnault, Amagat and others showed that at high pressures or low temperatures, considerable

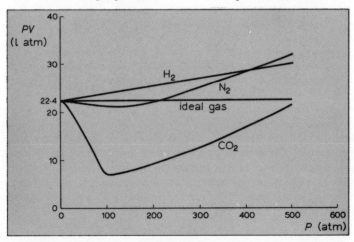

Fig. 2. Values of PV for one mole of gas at various pressures and at constant temperature

deviations from ideal behaviour occurred. Some examples are shown in Figure 2, and the reasons underlying this departure from ideality are discussed on page 37.

Charles' law (sometimes called Gay Lussac's law since the two workers arrived at the same conclusion independently) *At a constant pressure, the volume of a given mass of gas expands by 1/273 of its volume at 0°C for each 1°C rise in temperature.*

If V is the volume of a certain mass of gas at $t°C$ while V_0 is the volume of the same mass of gas at 0°C, then at a fixed pressure Charles' law is expressed by the relation

$$V = V_0 \left(1 + \frac{t}{273}\right)$$

The factor 1/273 is the coefficient of cubical expansion of a gas, and this value should be the same for all gases. Again, deviations from the ideal behaviour required by this equation are found with all gases, but at low pressures and normal temperatures most gases obey both Boyle's and Charles' laws. Under these conditions of near ideal behaviour, the coefficient of cubical expansion is 1/273·16. Taking this value as the coefficient of cubical expansion of an ideal gas and extrapolating back as shown in Figure 3, the volume of an

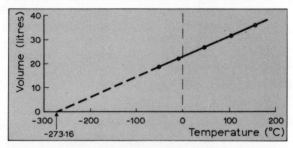

Fig. 3. Variation of the volume of a gas with temperature

ideal gas falls to zero at $-273·16°C$. This is the *absolute zero* of temperature. The scale of temperature in centigrade degrees, starting from absolute zero, is known as the *absolute scale*. Temperatures quoted on this scale are expressed in degrees Kelvin, K, and the use of this scale is implied by using the symbol T to represent temperature in an equation. Consequently, $t + 273·16 = T$ K.

Since

$$V = V_0 \left(1 + \frac{t}{273·16}\right)$$

then
$$V = \frac{V_0}{273 \cdot 16}(273 \cdot 16 + t)$$

For a given mass of gas at a fixed pressure, the fraction $V_0/273 \cdot 16$ is a constant, say k'. Hence
$$V = k'(273 \cdot 16 + t)$$
$$= k'T$$

which is an alternative form of Charles' law.

The Ideal Gas Equation

Boyle's and Charles' laws may be combined to form the ideal gas equation as follows: If

$$V \propto \frac{1}{P} \quad \text{at a fixed temperature}$$

and

$$V \propto T \quad \text{at a fixed pressure}$$

then

$$V \propto \frac{T}{P} \quad \text{when both temperature and pressure vary}$$

Rearranging

$$PV \propto T$$
$$PV = kT \quad \text{where } k \text{ is a constant}$$
$$\frac{PV}{T} = k$$

Alternatively, when the volume of a given mass of gas alters from V_1 to V_2 corresponding to changes in pressure from P_1 to P_2 and temperature from T_1 to T_2, the ideal gas equation can take the form

$$\frac{P_1 V_1}{T_1} = \frac{P_2 V_2}{T_2}$$

since each fraction is equal to the constant k.

Avogadro's hypothesis states that equal volumes of gases at the same temperature and pressure contain equal numbers of molecules. Taking the converse of this, one mole (i.e. N molecules) of any gas at the same temperature and pressure occupies the same volume, which means that the value of k in the ideal gas equation, $PV/T = k$ is the same for one mole of any gas assumed to behave ideally. For one mole of gas, the symbol R replaces k, and R is known as the *molar gas constant*. Thus

$$\frac{PV}{T} = R$$

or

$$PV = RT$$

This equation, which is a combination of Boyle's law, Charles' law and Avogadro's hypothesis, applies only to, and thus defines, an ideal gas. Consequently, it is referred to as the *ideal gas equation*.

For n moles of gas,

$$PV = nRT$$

The Gas Constant

The value of the molar gas constant, R, depends on the units in which P and V are measured. Pressure is defined as force per unit area or (force) × (length)$^{-2}$ while volume has the dimensions of (length)^3. The dimensions of the product PV are therefore (force) × (length) which are those of energy. Hence R, equal to PV/T, has to be expressed in units of (energy) (degree)$^{-1}$ (mole)$^{-1}$, and the energy units commonly used are (a) litre-atmospheres, (b) joules and (c) calories, although the latter is now becoming obsolete following the introduction of SI units (see p. 102).

After allowing for deviations from ideality, it has been estimated from measurements on real gases, that one mole of ideal gas should occupy 22·414 litres at s.t.p. (1 atmosphere pressure and 273·16 K). Hence

$$R = \frac{PV}{T} = \frac{1 \times 22\cdot414}{273\cdot16} = 0\cdot08205 \text{ l atm deg}^{-1} \text{ mole}^{-1}$$

To obtain the value of R in joules, the pressure must be expressed in c.g.s. units. A pressure of 1 atmosphere is the force per square centimetre experienced at the foot of a vertical column of mercury (density 13·595 g cm^{-3}) 76 cm high, under a gravitational force of 980·66 cm s^{-2}. Then

$$R = \frac{76 \times 13\cdot595 \times 980\cdot66 \times 22414}{273\cdot16}$$

$$= 8\cdot314 \times 10^7 \text{ erg deg}^{-1} \text{ mole}^{-1}$$

Since 1 joule = 10^7 erg,

$$R = 8\cdot314 \text{ J deg}^{-1} \text{ mole}^{-1}$$

(Also, 1 calorie = 4·184 J, and hence

$$R = \frac{8 \cdot 314}{4 \cdot 184} = 1 \cdot 987 \text{ cal deg}^{-1} \text{ mole}^{-1}$$

or very nearly 2 cal deg^{-1} mole^{-1}, a value which is frequently used in calculations.)

Calculations on the Gas Laws

Errors in these calculations usually arise from:

(a) using inconsistent units on opposite sides of the equation

$$\frac{P_1 V_1}{T_1} = \frac{P_2 V_2}{T_2}$$

(for example, if P_1 is given in atmospheres, P_2 must be in the same units),
(b) not converting temperatures to the absolute scale,
(c) either omitting, or using an incorrect value of n in the equation $PV = nRT$

Example 8 5 litres of air, measured at 27°C and 75 cm pressure are compressed to 5 atmospheres while the temperature rises to 43°C. Calculate the final volume of the air.

Solution

$$P_1 = 75 \text{ cm}$$
$$P_2 = 5 \times 76 \text{ cm}$$
$$T_1 = 27 + 273 = 300 \, K$$
$$T_2 = 43 + 273 = 316 \, K$$
$$V_1 = 5 \text{ litres}$$

Using

$$\frac{P_1 V_1}{T_1} = \frac{P_2 V_2}{T_2}$$

$$V_2 = \frac{P_1 V_1}{T_1} \times \frac{T_2}{P_2}$$

$$= \frac{75 \times 5 \times 316}{300 \times 5 \times 76} = 1 \cdot 04 \text{ litres.}$$

Example 9 What weight of oxygen would be required to fill a 2 litre flask at 20°C and 745 mm pressure if R has the value 0·082 l atm deg^{-1} mole^{-1}?

Solution Using $PV = nRT$,

$$n = \frac{PV}{RT}$$

Converting the pressure to atmospheres, $P = 745/760$. Hence

$$n = \frac{745}{760} \times \frac{2}{0.082 \times 293} = 0.0815 \text{ mole}$$

Since 1 mole of oxygen weighs 32·0 g, the weight required is

$$0.0815 \times 32 = 2.61 \text{ g}$$

Example 10 What is the pressure of a mixture of 1 g of hydrogen and 1·4 g of nitrogen stored in a 5-litre vessel at 127°C?

Solution

$$\text{Number of moles of hydrogen gas} = \frac{1}{2} = 0.5$$

$$\text{number of moles of nitrogen gas} = \frac{1.4}{28} = 0.05$$

$$\text{total number of moles of gas} = n = 0.55$$

Using $PV = nRT$,

$$P = \frac{0.55 \times 0.082 \times 400}{5}$$

$$= 3.61 \text{ atm}$$

An alternative method of solving Example 10 would be to calculate the pressure exerted by the hydrogen and nitrogen individually:

$$\text{pressure exerted by hydrogen, } p_{H_2} = \frac{0.5 \times 0.082 \times 400}{5}$$

$$= 3.284 \text{ atm}$$

$$\text{pressure exerted by nitrogen, } p_{N_2} = \frac{0.05 \times 0.082 \times 400}{5}$$

$$= 0.328 \text{ atm}$$

$$\text{total pressure, } P = p_{H_2} + p_{N_2} = 3.284 + 0.328$$

$$= 3.61 \text{ atm}$$

This is an example of the following law.

Dalton's law of partial pressures *For gases which do not react, the total pressure of a mixture of gases is the sum of the partial pressures of each constituent.*

The partial pressure of a gas is the pressure it would exert if it alone occupied the total volume. In the above example, p_{H_2} and p_{N_2} are the partial pressures of hydrogen and nitrogen respectively.

The partial pressure of a gas is proportional to the number of moles of gas present. This fact is important in calculations on steam distillations, as shown in Example 11.

Example 11 A mixture of nitrobenzene and water boils at 99°C to give a vapour in which the partial pressure of water is 733 mm and that of nitrobenzene is 27 mm. Calculate the ratio by weight of nitrobenzene to water in the condensate.

Solution The number of moles of each constituent is proportional to its partial pressure. Thus,

$$n_{H_2O} \propto p_{H_2O} \quad \text{and} \quad n_{nitrobenzene} \propto p_{nitrobenzene}$$

On dividing

$$\frac{n_{nitrobenzene}}{n_{H_2O}} = \frac{p_{nitrobenzene}}{p_{H_2O}} = \frac{27}{733}$$

Hence,

$$\frac{\text{weight of nitrobenzene}}{\text{weight of water}} = \frac{27 \times 133}{733 \times 18} = \frac{1}{398}$$

The law of partial pressures is important in connection with the collection of gases over water. Since gases collected this way always contain water vapour, the vapour pressure of water (at the particular temperature) is subtracted from the total pressure to give the true pressure of the gas collected. For example, the partial pressure of a gas collected over water at 760 mm and 15°C is $760 - 13 = 747$ mm since the vapour pressure of water at 15°C is 13 mm.

Example 12 A 2-litre flask containing carbon dioxide at 60 cm pressure is connected to a 4-litre flask containing nitrogen at 48 cm pressure. If the temperature is kept constant, calculate the final pressure of the mixture.

Solution As an alternative to using a repeated application of Boyle's law, we can proceed as follows:

The final total volume occupied by the gases is 6 litres; thus, the partial pressure of carbon dioxide falls to 2/6 of its initial value, that is,

$$p_{CO_2} = \tfrac{2}{6} \times 60 = 20 \text{ cm}$$

Similarly,
$$p_{N_2} = \tfrac{4}{6} \times 48 = 32 \text{ cm}$$
Hence the final pressure is 52 cm.

Dalton's Law and Real Gases

Dalton's law applies only to ideal gases (see p. 24) and for real gases the total pressure is usually slightly in excess of the sum of the partial pressures of the constituents. At very high pressures, intermolecular attraction becomes significant and this may bring about a negative deviation. At low pressures (up to a few atmospheres) and normal temperatures, the law may be regarded as fairly accurate.

Mole Fraction

The mole fraction x of a species in a mixture is the ratio of the number of moles of that species to the total number of moles present. For example, in a mixture of three components, A, B and C, the mole fraction of A is

$$x_A = \frac{n_A}{n_A + n_B + n_C}$$

where n_A, n_B and n_C are the number of moles of each constituent present.

Example 13 What is the mole fraction of nitrogen in a mixture of nitrogen and hydrogen in which the partial pressure of hydrogen is 63 cm and the total pressure is 90 cm?

Solution The partial pressure of nitrogen is
$$90 - 63 = 27 \text{ cm}$$
Hence,
$$x_{N_2} = \frac{n_{N_2}}{n_{N_2} + n_{H_2}} = \frac{p_{N_2}}{p_{N_2} + p_{H_2}} = \frac{27}{90} = 0\cdot3$$

The Kinetic Theory of Gases

This theory interprets ideal gas behaviour in terms of molecular motion, and in consequence, the reasons for the behaviour of real gases can be appreciated. The basis of the theory for an ideal gas may be summarized as follows:

1. An ideal gas consists of a large number of particles (representing molecules) moving at random.

2. Each particle is regarded as a point, i.e. the size of each particle is infinitesimal and the total volume occupied by all the particles is insignificant.
3. The particles are perfectly elastic and no energy is lost when they collide.
4. No forces of attraction obtain between the particles.
5. The particles are in frequent collision with each other and with the walls of the containing vessel. The latter collisions produce the measured pressure exerted by the gas. Between collisions, the particles move in straight lines.
6. A rise in temperature will result in a faster movement of the particles, thus producing an increase in pressure if the volume remains constant.
7. If the volume of the container is decreased, the particles will collide with the walls more frequently and an increase in pressure will result.

Derivation of the Fundamental Equation for the Pressure of a Gas

A cube-shaped vessel of side length l cm and of volume V cm^3 contains one mole of ideal gas (represented by N particles each of mass m) at a pressure P and at a fixed temperature. The behaviour of the particles is in accordance with the assumptions detailed above.

Taking one particle moving at random, its velocity u may be resolved into three components mutually at right angles such that

$$u^2 = u_x^2 + u_y^2 + u_z^2$$

where u_x is the component of velocity along the x-axis, etc. When the container is placed with one corner at the origin of axes, the sides of the vessel can be made to coincide with the axes as shown in Figure 4.

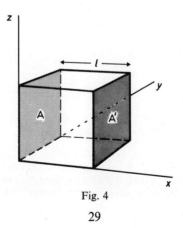

Fig. 4

Considering the direction of the x-axis, the particle moves with velocity u_x between the faces A and A' which are l cm apart, and the time taken for the particle to travel between A and A' is l/u_x.

The number of times per second that the particle collides with *either* A *or* A' will be u_x/l.

The particle approaches each face with momentum mu_x and leaves with a momentum $-mu_x$ (or vice versa), since the collisions are perfectly elastic. The change of momentum accompanying each collision is thus $2mu_x$, and the rate of change of momentum per second is

$$2mu_x \frac{u_x}{l} = \frac{2mu_x^2}{l}$$

This is the force on A and A' taken together. Similarly, the force exerted by this particle on all six faces of the vessel is

$$\frac{2m}{l}(u_x^2 + u_y^2 + u_z^2)$$

and the total force due to N particles is

$$\frac{2mN}{l}(\overline{u_x^2} + \overline{u_y^2} + \overline{u_z^2})*$$

or

$$\frac{2mN}{l}(\overline{u^2})$$

Since pressure is defined as force per unit area, and the total area of the six faces of the container is $6l^2$, the pressure of the gas is

$$P = \frac{2mN\overline{u^2}}{l} \times \frac{1}{6l^2}$$

$$= 1m\frac{N\overline{u^2}}{3l^3}$$

Putting $l^3 = V$, the fundamental gas equation results:

$$PV = \tfrac{1}{3}mN\overline{u^2}$$

Energy of the Particles

The system considered in connection with the derivation of the

* $\overline{u_x^2}$ represents the average of the squares of the velocity components in the x-direction for N particles, i.e. it is a mean square velocity. Note that the mean velocity u_x is zero, since in a large assembly the number of particles moving from A to A' $(+u)$ equals the number moving in the reverse direction $(-u)$.

fundamental equation was assumed to be an ideal gas consisting of minute particles moving at random. The energy of these particles is kinetic energy of movement within the gas volume (translational energy). Since kinetic energy is reckoned as $\tfrac{1}{2}mu^2$, the quantity $\tfrac{1}{2}mN\overline{u^2}$ is equal to the kinetic energy of all the particles in one mole of gas. Thus

$$PV = \tfrac{2}{3} \cdot \tfrac{1}{2}mN\overline{u^2} = \tfrac{2}{3}E_k$$

where E_k is the total kinetic energy of the particles in the gas. Heating the gas brings about a rise in temperature corresponding to a conversion of heat energy into kinetic energy. Consequently, if the temperature remains unaltered, the value of E_k is unchanged which is in agreement with Boyle's law that PV is constant.

The particles in a gas cannot all have the same velocity even at a fixed temperature on account of the many collisions that take place. However, J. Clerk Maxwell (1860) deduced that the mean kinetic energy per particle, $\tfrac{1}{2}\overline{mu^2}$ is the same for all particles irrespective of mass at the same temperature.

For equal volumes of two gases A and B at the same pressure,

$$PV = \tfrac{1}{3}N_A m_A \overline{u_A^2} = \tfrac{1}{3}N_B m_B \overline{u_B^2}$$

or

$$N_A m_A \overline{u_A^2} = N_B m_B \overline{u_B^2}$$

If these gases are kept at the same temperatures, Maxwell's deduction demands that

$$\tfrac{1}{2}m_A \overline{u_A^2} = \tfrac{1}{2}m_B \overline{u_B^2}$$

from which it follows that $N_A = N_B$. This conclusion, that equal volumes of gases at the same temperature and pressure contain equal numbers of molecules, is, of course, Avogadro's hypothesis.

So far, the kinetic theory has been discussed in terms of particles rather than molecules. This distinction has been made to emphasize the fact that the actual behaviour of molecules in real gases will deviate from the assumed behaviour of particles constituting an ideal gas. This point is taken up again on page 37.

Temperature and Molecular Velocities

For one mole of gas,

$$PV = RT$$

and from the kinetic theory

$$PV = \tfrac{1}{3}mN\overline{u^2}$$
$$= \tfrac{1}{3}M\overline{u^2} = \tfrac{2}{3}E_k$$

where M represents the weight of one mole of gas.

By comparing these equations, it follows that

$$\tfrac{1}{3}M\overline{u^2} = \tfrac{2}{3}E_k = RT$$

From this equation, it is clear that

(a) the kinetic energy of a gas is proportional to the absolute temperature,
(b) the molecular velocity is proportional to the square root of the absolute temperature, and depends also on the molar weight of the gas.

The equation $\tfrac{1}{3}M\overline{u^2} = RT$ can be re-written to give

$$\overline{u^2} = \frac{3RT}{M}$$

or,

$$\sqrt{(\overline{u^2})} = \sqrt{\frac{3RT}{M}}$$

$\sqrt{(\overline{u^2})}$ is called the root mean square velocity, and for hydrogen at s.t.p. its value is

$$\sqrt{(\overline{u^2})} = \sqrt{\frac{3RT}{M}} = \sqrt{\frac{3 \times 8\cdot31 \times 10^7 \times 273}{2\cdot016}}$$

$$= 184\,000 \text{ cm s}^{-1}$$

Some other root mean square velocities at s.t.p. are

oxygen, $46\,140$ cm s^{-1} carbon dioxide, $39\,300$ cm s^{-1}
helium, $130\,600$ cm s^{-1} chlorine, $30\,950$ cm s^{-1}

It can be shown that the average velocity of the molecules $\bar{u}$ is slightly less than the root mean square velocity. The two quantities are related by the equation $\bar{u} = 0\cdot92 \sqrt{(\overline{u^2})}$.

The following method for the determination of molecular velocities outlines the chief techniques used by a number of workers investigating this problem. Two discs, each pierced by a slit, are set some distance apart on a common axle. The slits are collinear, and a beam of molecules is directed on to the first disc, while a detecting device is placed behind the second disc. When the discs are stationary, the beam of molecules passes through the two slits and is recorded by the detector. On rotation of the discs, molecules will again enter the detector when the time taken by a molecule to travel between the discs is equal to the time required for the discs to complete one revolution. As a result of these investigations, the majority of molecules were found to have average velocities equal to the values calculated from the kinetic theory. Some scatter of velocities about the mean value was also apparent. Such a scatter is expected, since the molecules are continually colliding and cannot have a steady velocity. The *distribution of velocities* was expressed in mathematical terms by J. Clerk Maxwell, in 1860, who showed that in a system containing N molecules, the number n of molecules having a kinetic energy greater than a particular value E, is given by

$$n/N = e^{-E/RT*}$$

This relation is expressed graphically in Figure 5, and it is clear that most molecules have a velocity close to the mean value.

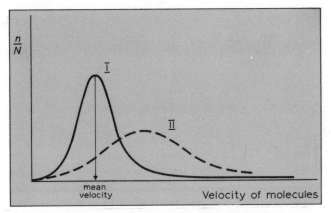

Fig. 5. Distribution of molecular velocities

Curve I represents the distribution of energies (or velocities since $E = \tfrac{1}{2}M\overline{u^2}$) at a lower temperature, while curve II relates to a higher temperature. From these curves, it can be seen that an increase in temperature causes a general increase in molecular velocities, but a greater scatter of velocities about the mean accompanies this increase. The results of the experimental determinations of molecular velocities agree well with the Maxwell distribution law.

Effusion and Diffusion

The root mean square velocity of a molecule of ideal gas is given by

$$\sqrt{(\overline{u^2})} = \sqrt{\frac{3RT}{M}}$$

* e, which has the value 2·71828..., is the base of natural or Naperian logarithms.

For two gases A and B at the same temperature,

$$\frac{\sqrt{(\overline{u_A^2})}}{\sqrt{(\overline{u_B^2})}} = \sqrt{\frac{3RT}{M_A} \times \frac{M_B}{3RT}} = \sqrt{\frac{M_B}{M_A}}$$

where M_A and M_B represent the weights of one mole of A and B respectively.

The passage of a gas through a fine hole in the wall of a vessel is known as *effusion* and the rate of escape of gas should, from the above equations (for the ideal case) be inversely proportional to its mole weight. T. Graham had arrived at this conclusion in 1833, following his experiments on gaseous effusion and diffusion.

Graham's law *The rates of effusion (or diffusion) of different gases (under identical conditions) are inversely proportional to the square roots of their densities (or mole weights).* For two gases A and B of densities d_A and d_B,

$$\frac{\text{rate of effusion of A}}{\text{rate of effusion of B}} = \frac{\sqrt{d_B}}{\sqrt{d_A}} = \frac{\sqrt{M_B}}{\sqrt{M_A}}$$

(Since 1 mole of gas, i.e. M grammes, occupies 22·4 l at s.t.p. then the density d of a gas at s.t.p. is

$$d = \frac{M}{22\cdot 4} \text{ g l}^{-1}$$

Hence, density may be replaced by the mole weight of the gas.)

The rate of effusion is the volume of gas escaping in unit time. The *time* taken for a given volume of gas to effuse is *inversely proportional* to the *rate* of effusion. Hence

$$\frac{\text{time taken for 1 volume of A to escape}}{\text{time taken for 1 volume of B to escape}} = \frac{\text{rate of effusion of B}}{\text{rate of effusion of A}}$$

$$= \frac{\sqrt{d_A}}{\sqrt{d_B}} = \frac{\sqrt{M_A}}{\sqrt{M_B}}$$

Example 14 The density of methane is four times that of helium. Calculate the time taken for 10 cm^3 of methane to escape through an orifice if, under identical conditions, helium effuses at the rate of 1·5 cm^3 per minute.

Solution If 1·5 cm^3 of helium effuse in 1 minute
then 10 cm^3 of helium would effuse in $1 \times 10/1\cdot 5$ minute

Let t be the number of minutes taken for 10 cm³ of methane to effuse. Then

$$\frac{t}{10/1 \cdot 5} = \frac{\sqrt{4}}{\sqrt{1}} = 2$$

$$t = \frac{2 \times 10}{1 \cdot 5} = 13 \cdot 3 \text{ minute.}$$

The effusiometer This instrument enables the times required by different gases to escape, under identical conditions, through a small hole in a metal plate to be compared. A diagram of the instrument is shown in Figure 6.

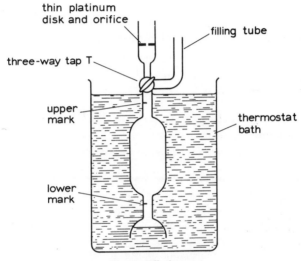

Fig. 6. Effusiometer

The apparatus is filled with the gas under test using the three-way tap, and placed in a thermostatically controlled bath. When a steady temperature has been obtained, the three-way tap T is adjusted to allow the gas under test to escape through the fine hole in the thin platinum disc at the upper end of the apparatus. The time required for the liquid to rise from the lower to the upper mark is noted. The experiment is repeated using a different gas at the same temperature. The liquid used in the thermostat bath should not, of course, react with or dissolve either gas.

Example 15 The time taken for a sample of hydrogen containing a small proportion of deuterium to escape from an effusiometer was 3·7 min. If, under the same conditions, an equal volume of pure oxygen effused in 14·1 min, calculate the proportion of deuterium in the sample. (Take the atomic weights of oxygen, hydrogen and deuterium as 16·0, 1·0 and 2·0 respectively.)

Solution

Thus
$$\frac{\text{Time for effusion of oxygen}}{\text{Time for effusion of hydrogen}} = \frac{\sqrt{M_{\text{oxygen}}}}{\sqrt{M_{\text{hydrogen}}}}$$

$$\frac{14\cdot1}{3\cdot7} = \frac{\sqrt{32}}{\sqrt{M_{\text{hydrogen}}}}$$

$$M_{\text{hydrogen}} = 2\cdot20$$

Let $x\%$ of deuterium be present in the hydrogen sample, then the average weight of one mole of hydrogen will be

$$\frac{x \times 4 + (100-x) \times 2}{100} = 2\cdot20$$

$$x = 10\%$$

In effusion, a gas is made to pass through a small orifice under pressure. Diffusion relates to the mixing of two gases at the same pressure. In spite of the fact that gas molecules have high velocities, diffusion is a slow process. This is due to the fact that each molecule is, at ordinary pressures, involved continually in colliding with other molecules, thus greatly restricting the rate of forward movement. However, the rate of bulk motion of a gas does depend on molecular velocities (and also on the mean free path of the molecules) which agrees with Graham's law, that the rate of diffusion is inversely proportional to the square root of the density of the gas. Since the rates of effusion and diffusion depend on gas densities, both methods are employed in the separation of gaseous isotopes.

Mean Free Path

The average distance traversed by a gas molecule between successive collisions is termed its *mean free path*. This quantity is inversely proportional to

   (a) the number of molecules per unit volume,
   (b) the square of the diameter of the molecule.

An estimation of molecular diameters can be made by measuring rates of diffusion which are related to the mean free path. At s.t.p.

the mean free path of a molecule in a gas is of the order of 10^{-5} cm. At low pressures, the number of molecules per unit volume is much less, and at 10^{-5} atm the mean free path is about 1 cm.

Real Gases and the Kinetic Theory

The kinetic theory explains the behaviour of ideal gases, but modifications of the theory are required to allow for the deviations from ideality shown by real gases. Two of the assumptions made at the outset are clearly untrue for real gases. These are:

(a) the gas molecules do not attract each other,
(b) the gas molecules occupy negligible volume.

Intermolecular Attraction If the gas molecules do attract each other, the gas will be more compressible than the ideal case. The effect of intermolecular attraction is to increase the effective pressure on the gas above the value indicated as the measured external pressure. Consequently, the volume will be less than expected and the value of PV will fall below the ideal case (as shown in the early parts of the curves for carbon dioxide and nitrogen in Figure 2, p. 21). Moreover, gases can be liquefied, and the liquid state implies a considerable degree of attraction between the molecules; hence it would be reasonable to expect that some degree of attraction between molecules continues in the gaseous state. It also follows that those gases in which the intermolecular attractions are greatest are most easily liquefied. When a gas expands through a porous plug, or an orifice, from a region of high to low pressure, a cooling effect, known as the *Joule–Thomson effect* is noticed. This cooling effect is due to the internal work performed in separating the molecules against their attractions. Hydrogen and helium behave exceptionally as they exhibit the Joule–Thomson effect only at low temperatures. Hydrogen, for example, cools on Joule–Thomson expansion below $-80°C$ at 120 atm, but warms slightly when expanded above this temperature.

The Volume of the Molecules

At low pressures, the volume occupied by the gas molecules is not significant, but at higher pressures the molecules are compacted to a degree such that they tend to behave like a liquid, which is almost incompressible. As a result, at higher pressures the volume of the gas remains larger than the ideal case, and in consequence the product PV rises. This is shown in the later part of the curves for carbon dioxide and nitrogen in Figure 2, p. 21. Since hydrogen shows an upward trend from ideal behaviour at all pressures, it is concluded that the effect of the volume of the molecules is greater than the effect of intermolecular attraction. In other words, the force of intermolecular attraction between gaseous hydrogen molecules is very small.

The van der Waals Equation

The effect of molecular size and intermolecular forces of attraction are taken into account in the van der Waals equation, in which the ideal gas equation $PV = RT$ is changed to

$$(P + a/V^2)(V - b) = RT$$

for one mole of ideal gas. It has already been mentioned (p. 37) that due to the forces of attraction between molecules, the actual pressure experienced by the gas molecules is greater than the indicated external pressure P. The first term in the van der Waals equation is to allow for this factor, and the fraction a/V^2 is arrived at by considering molecules such as B in Figure 7, which are just

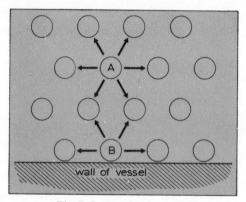

Fig. 7. Intermolecular attraction

about to strike the wall of the container. Such molecules experience an unbalanced force of attraction due to molecules such as A in the bulk of the gas. The force with which a molecule strikes the wall is lessened (i.e. the measured pressure is too low) on account of these unbalanced forces.

Now the loss in pressure due to the inward force of attraction is proportional to the *force of attraction per molecule* multiplied by the *number of molecules moving to the wall*, and each of these quantities is proportional to the *number of molecules per unit volume*. Hence

$$\text{loss in pressure} \propto \frac{n}{V} \times \frac{n}{V} = \frac{a}{V^2}$$

where a is a constant and n is the total number of molecules in the volume V. The term a/V^2 which allows for the intermolecular force of attraction is called the *internal pressure*. The second term in the van der Waals equation, $(V-b)$, contains a correction factor b which is related to the actual volume occupied by the molecules.

The values of a are expressed in l^2 atm while b is given in litres. Some typical values are:

hydrogen	$a = 0.25$	$b = 0.0267$
oxygen	$a = 1.52$	$b = 0.0312$
carbon dioxide	$a = 3.60$	$b = 0.0428$
sulphur dioxide	$a = 6.7$	$b = 0.056$

The increasing value of a agrees with increasing ease of liquefaction from hydrogen to sulphur dioxide.

Example 16 One mole of carbon dioxide is found to occupy 0·382 litres at 40°C and under a pressure of 50 atm. Calculate the pressure that the gas would be expected to exert using (a) the ideal gas equation, (b) the van der Waals equation. Use the values of a and b given above, and take R as 0·082 l atm deg^{-1} mole^{-1}.

Solution (a) Using $PV = RT$

$$P = \frac{0.082 \times 313}{0.382} = 67.2 \text{ atm}$$

(b) Using

$$(P + a/V^2)(V - b) = RT$$

and rearranging,

$$P = \frac{RT}{(V-b)} - \frac{a}{V^2}$$

$$= \frac{0.082 \times 313}{(0.382 - 0.0428)} - \frac{3.6}{(0.382)^2}$$

$$= 50.9 \text{ atm}$$

Other Equations of State

Although the van der Waals equation agrees much more closely with the behaviour of real gases than the ideal gas equation, it is found that the values of a and b are not constant but vary with temperature. A number of alternative equations of state have been proposed, such as that of Berthelot:

$$(P+a'/TV^2)(V-b) = RT$$

which gives good results at moderate pressures.

Liquefaction of Gases Under Pressure

Following the work of Faraday and others, some gases, such as hydrogen chloride, sulphur dioxide, ammonia and chlorine were found to change to the liquid state under the influence of increased pressure. Other gases, even when subjected to pressures of 3000 atmospheres, failed to liquefy, and they were regarded as permanent gases, i.e. they could not be liquefied. The experiments of Andrews showed that liquefaction of a gas is not possible if the temperature is above the critical value.

Andrews' Isothermals for Carbon Dioxide

The results of Andrews' investigations on the pressure-volume relations at high pressures and at various temperatures are shown graphically in Figure 8. Each curve is an isothermal, that is, the

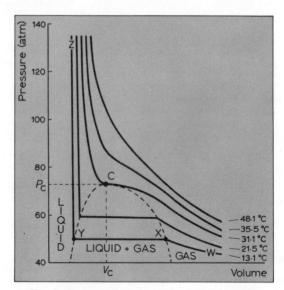

Fig. 8. Andrews' isothermals for carbon dioxide

pressure-volume relationship expressed by a given curve applies only to the temperature shown. If carbon dioxide obeys Boyle's law, a graph of pressure against volume should produce a rectangular hyperbola, and at 48·1°C this is found to be the case. At successively lower temperatures, the isothermals show increasing deviations from the expected curve. The isothermal at 13·1°C can be divided into three distinct parts:

(a) Along WX, the volume of the gas decreases corresponding to an increase in pressure.
(b) Between X and Y, the volume decreases rapidly following a slight increase in pressure. Along this line, the gas is liquefying and at Y, the carbon dioxide is completely liquid.
(c) The portion YZ shows the very slight change experienced by a liquid under increasing pressure.

The approximately parabolic dotted line joins the ends of the horizontal portions analogous to XY, and at C, these horizontal portions disappear altogether. The point C is known as the *critical point* for the gas, and three critical quantities are defined:

(a) the *critical temperature* (T_C) is the temperature above which a gas fails to liquefy under compression;
(b) the *critical pressure* (P_C) is the pressure which is just sufficient to liquefy the gas at its critical temperature;
(c) the *critical volume* (V_C) of a gas is the volume of one mole of gas at the critical temperature and pressure.

All gases show a critical point, and the values of T_C, P_C and V_C are peculiar to each gas. For example, the critical temperature for carbon dioxide is 304·2 K, and the critical pressure is 72·9 atm, while the corresponding values for oxygen are 154·3 K and 49·7 atm.

Usually, when a liquid changes into a gas, a large increase in volume is noted since the gas has a much smaller density than the liquid. As the critical point is approached, the densities of both the liquid and the gas become equal, which implies that there is no difference between a liquid and a gas under these conditions. It is found that no sudden changes in volume or density occur when a liquid changes into a gas near the critical point, but the liquid meniscus suddenly vanishes. The term *continuity of state* is used to describe this smooth transition from one phase to another.

On compressing a gas (below its critical temperature) in a very clean apparatus with smooth inner surfaces, it is possible to hold the gas at a pressure above that at which liquefaction normally takes place. This is shown by the dotted line XA in Figure 9, in which WXYZ is the normal isothermal.

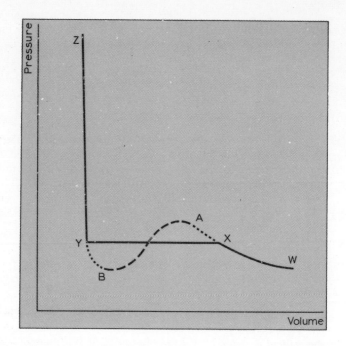

Fig. 9

An increase in pressure beyond A results in a rapid liquefaction of the gas. In a similar way, if the pressure on a pure liquid is smoothly and gradually reduced, the pressure–volume relation follows the dotted curve YB. It has been suggested that joining A and B (by the broken curve shown) would produce a pressure–volume relation relevant to that for an ideal gas.

Now the van der Waals equation can be re-written in the form of a cubic equation:

$$\left(P + \frac{a}{V^2}\right)(V-b) = RT$$

$$V^3 - \left(b + \frac{RT}{P}\right)V^2 + \frac{aV}{P} - \frac{ab}{P} = 0$$

and this agrees with the idea of continuity of state and the shape of the ideal isotherm WABZ. Moreover, a cubic equation has three real roots (corresponding to isotherms below the critical point), a point of inflexion (at the critical point) or one real root and two imaginary roots (corresponding to the gaseous state).

Liquefaction of Gases

Cooling a gas held under a high pressure is the basis of many methods used for liquefying gases. Pictet used a two-stage process to cool oxygen below its critical temperature of $-119°C$. Modern methods employ the Joule–Thomson effect or adiabatic expansion to reduce the temperature of the gas below its critical value. The Joule–Thomson effect has already been mentioned, while the adiabatic expansion method relies on the fact that an expanding gas doing work against an external pressure cools down, if the system is insulated against heat gain or loss. The relation between pressure and temperature in an adiabatic expansion is

$$\left(\frac{T_1}{T_2}\right)^\gamma = \left(\frac{P_1}{P_2}\right)^{\gamma-1}$$

where γ is the ratio of the molar heat capacities of the gas at constant pressure and at constant volume. Consequently, the higher the initial pressure, the greater the fall in temperature on expansion.

The Joule–Thomson effect has been used for the production of liquid air since 1895 when Linde and Hampson both used this method for obtaining low temperatures. For every gas there is an inversion temperature above which the Joule–Thomson effect produces heating, and below which cooling occurs. With some gases, such as hydrogen, pre-cooling is necessary before the desired results can be obtained.

In the modern process for the production of liquid air, compressed air is bubbled through concentrated sodium hydroxide solution (to remove carbon dioxide) and dried over activated alumina. The air is then allowed to expand through a fine nozzle, and the cooled air is passed through a counter-current heat exchanger against the incoming high pressure stream. After a short time, expansion results in the formation of liquid air which may be stored in vacuum vessels, or fractionated to give oxygen, nitrogen and argon.

Relative (or Vapour) Densities of Gases

The *relative density* of a gas is defined as *the ratio of the weights of equal volumes of the gas and hydrogen, measured under identical conditions*. From this definition, the relative density of a gas is given by:

$$\text{R.D. (or V.D.)} = \frac{\text{weight of one volume of gas}}{\text{weight of one volume of hydrogen}}$$

$$= \frac{\text{weight of } n \text{ molecules of gas}}{\text{weight of } n \text{ molecules of hydrogen}} \quad \text{(from Avogadro's hypothesis)}$$

Thus

$$\text{R.D.} = \frac{\text{weight of 1 mole of gas}}{\text{weight of 1 mole of hydrogen}}$$

$$= \frac{\text{weight of 1 mole of gas}}{2}$$

Therefore, the weight of 1 mole of gas is twice the relative density:

$$\text{weight of 1 mole of gas} = 2 \times \text{R.D.}$$

Example 17 The densities of oxygen, helium and hydrogen at s.t.p. are 1·427, 0·1784 and 0·0892 g l^{-1} respectively. Calculate the weights of 1 mole of oxygen and helium.

Solution

$$\text{R.D. of oxygen} = \frac{1\cdot427}{0\cdot0892} = 16\cdot0$$

Thus

$$\text{1 mole of oxygen weighs } 2 \times 16\cdot0 = 32\cdot0 \text{ g}$$

Similarly

$$\text{R.D. of helium} = \frac{0\cdot1784}{0\cdot0892} = 2\cdot00$$

Thus

$$\text{1 mole of helium weighs } 2 \times 2\cdot00 = 4\cdot00 \text{ g}$$

Determination of the Approximate Weights of One Mole of Gas

The following methods are based on measuring the density of a gas or vapour.

Regnault's method (for gases)

1. A flask is filled with distilled water at a known temperature and weighed. From this weighing, the volume of the flask can be calculated.
2. The flask is emptied, dried, evacuated and weighed again after filling with gas at a known temperature and pressure.
3. The density of the gas is calculated, and the weight of one mole of gas is found.

Example 18 A flask weighs 129·1427 g when evacuated, and 442·8457 g when filled with water at 4°C. When filled with oxygen at 15·3°C and 754·2 mm pressure, the flask weighs 129·5647 g.

Calculate the weight of one mole of oxygen if the density of hydrogen is $0{\cdot}09$ g l^{-1} at s.t.p.

Solution

Weight of flask and water at 4°C = 442·8457 g
Weight of evacuated flask = 129·1427 g
Volume of flask = $\overline{313{\cdot}7030}$ cm^3

(since the density of water is 1·00 at 4°C). The volume of oxygen is 313·703 cm^3 at 15·3°C and 754·2 mm pressure. Therefore,

$$\text{volume of oxygen at s.t.p.} = \frac{313{\cdot}703 \times 754{\cdot}2 \times 273}{288{\cdot}3 \times 760}$$

$$= 294{\cdot}8 \text{ cm}^3$$

Now 1000 cm^3 hydrogen at s.t.p. weighs 0·09 g, so

$$\text{weight of } 294{\cdot}8 \text{ cm}^3 \text{ hydrogen} = \frac{0{\cdot}09 \times 294{\cdot}8}{1000} = 0{\cdot}02653 \text{ g}$$

Hence,

$$\text{R.D. of oxygen} = \frac{129{\cdot}5647 - 129{\cdot}1427}{0{\cdot}02653}$$

$$= \frac{0{\cdot}4220}{0{\cdot}02653} = 15{\cdot}91$$

Thus, the weight of one mole of oxygen is $2 \times 15{\cdot}91 = 31{\cdot}8$ g.

The accuracy of Regnault's method can be improved by correcting for buoyancy effects.

Dumas' method In this method, the density of the vapour produced by the evaporation of a volatile liquid is measured. The vapour is contained in a Dumas bulb which is surrounded by a bath of hot liquid kept at a steady temperature which is above the boiling point of the liquid under investigation. The experiment is carried out in the following stages:

1. The Dumas bulb is weighed full of air.
2. A small quantity of the liquid under test is introduced into the bulb.
3. The bulb is maintained at a steady temperature in the heating bath until all the liquid has vaporized and the vapour fills the bulb at atmospheric pressure. The excess vapour escapes into the atmosphere.

4. The tip of the bulb is then sealed by application of a hot flame.
5. The bulb is allowed to cool to room temperature and weighed.
6. The volume of the bulb is found either by weighing when filled with water, or by measuring the volume of water held by the bulb.

This experiment may be performed at high temperatures using quartz bulbs and liquids of high boiling points (for example, molten metals) in the heating bath. It is also possible to obtain good results with small scale apparatus.

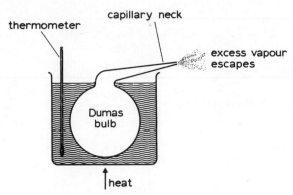

Fig. 10. Dumas apparatus

Example 19 A small scale Dumas bulb weighs 6·1563 g when full of air and 6·2097 g after filling with tetrachloromethane vapour and sealing. The volume of the bulb is 41·8 cm³ and the temperature of the heating bath is 93·0°C. Calculate the weight of one mole of tetrachloromethane vapour if the laboratory temperature is 20°C and the atmospheric pressure is 750 mm, given that the density of air is $1·293 \times 10^{-3}$ g cm^{-3} at s.t.p.

Solution

$$\text{Volume of air in bulb corrected to s.t.p.} = \frac{41·8 \times 750 \times 273}{293 \times 760}$$

$$= 38·44 \text{ cm}^3$$

weight of air in bulb = $38·44 \times 1·293 \times 10^{-3}$

$$= 0·0497 \text{ g}$$

Therefore

true weight of bulb = 6·1563 − 0·0497 = 6·1066 g

Also,
$$\text{weight of tetrachloromethane vapour} = 6{\cdot}2097 - 6{\cdot}1066$$
$$= 0{\cdot}1031 \text{ g}$$

This weight of vapour occupies $41{\cdot}8$ cm^3 at 750 mm and 93°C. Thus,
$$\text{volume of vapour corrected to s.t.p.} = \frac{41{\cdot}8 \times 750 \times 273}{366 \times 760}$$
$$= 30{\cdot}78 \text{ cm}^3$$

If $30{\cdot}78$ cm^3 vapour weighs $0{\cdot}1031$ g, then
$$\text{weight of 22 414 cm}^3 \text{ vapour} = \frac{0{\cdot}1031 \times 22\,414}{30{\cdot}78} = 75{\cdot}1 \text{ g}$$

Hence, the weight of one mole of tetrachloromethane is $75{\cdot}1$ g. (The correct result is in fact $76{\cdot}9$ g.)

Victor Meyer's method This is another method which measures the volume occupied by a given weight of vapour produced on the evaporation of a volatile liquid. The apparatus is shown in Figure 11.

1. About $0{\cdot}1$ g of the volatile liquid is accurately weighed out in a Hofmann bottle.
2. The Hofmann bottle (lightly stoppered) is allowed to rest on a glass rod in the upper part of the apparatus as shown.
3. The heating jacket is raised to a temperature in excess of the boiling point of the liquid under investigation.
4. When a steady state has been reached, no bubbles of air will be displaced by expansion through the side tube, and at this stage the graduated air receiver is placed in the position shown and the glass rod is withdrawn sufficiently to allow the Hofmann bottle to fall into the bulb B.
5. The stopper of the Hofmann bottle is forced out as the liquid vaporizes. (A jammed stopper is a frequent cause of failure in this experiment.) The vapour produced displaces an equivalent quantity of air into the measuring receiver.
6. The final volume of air displaced is read off, the temperature of the water in the air receiver is noted and the barometric pressure is read. The pressure of air in the measuring receiver is less than atmospheric by an amount equal to the hydrostatic head of liquid remaining in the receiver, and a correction must also be applied for the presence of water vapour in the displaced air.

This method is also capable of being performed at high temperatures and on a small scale.

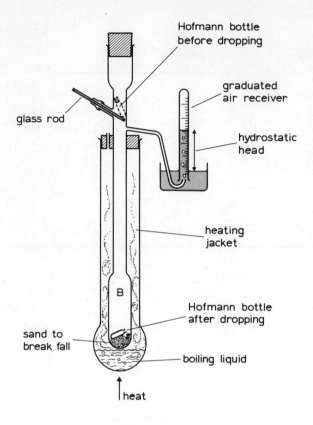

Fig. 11. Victor Meyer apparatus

Example 20 Calculate the weight of one mole of chloroform from the following results obtained by Victor Meyer's method.

Weight of sample taken = 0·0884 g
Volume of air displaced = 18·4 cm^3 at 15·3°C and 762 mm pressure
Head of water remaining in measuring receiver = 9·8 cm

(The vapour pressure of water at 15°C is 13 mm.)

Solution The partial pressure of air in the measuring receiver is

$$762 - \frac{98}{13 \cdot 59} - 13 = 742 \text{ mm}$$

The volume of air displaced at s.t.p. (which is also the volume of vapour formed from the liquid sample, corrected to s.t.p.) is

$$\frac{18\cdot 4 \times 742 \times 273}{288\cdot 3 \times 760} = 17\cdot 01 \text{ cm}^3$$

If $17\cdot 01$ cm^3 vapour weighs $0\cdot 0884$ g, then 22414 cm^3 vapour weighs

$$\frac{0\cdot 0884 \times 22414}{17\cdot 01} = 116\cdot 5 \text{ g}$$

Thus, one mole of chloroform weighs $116\cdot 5$ g.
(The correct result is in fact $119\cdot 5$ g.)

The Gas Density Microbalance

This method, used by R. W. Gray and W. Ramsay (1910) to determine the density of radon, is capable of giving very accurate results using very small quantities of gas. The apparatus is shown diagrammatically in Figure 12. The bulb A, of about 8 cm^3 capacity, is

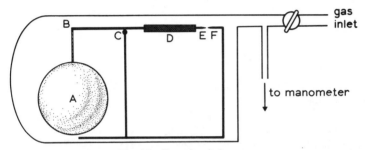

Fig. 12. Gas microbalance

suspended by a quartz fibre from the balance arm BC, which pivots at C on a transverse quartz fibre. D is a counter-balance plate whose surface area is equal to that of the bulb A. The plate D carries a pointer E which is used in conjunction with a second pointer F attached to the frame (not shown in full) on which the balance is mounted. This part of the apparatus is made entirely of quartz. The balance is enclosed in a glass cylinder, which can be filled with gas at any desired pressure, while the complete unit is maintained at a steady temperature. The apparatus is evacuated and filled with a gas (such as air or nitrogen) of known density. The buoyancy effect of the gas causes the bulb A to rise and the pressure of the gas is adjusted until the pointers E and F (viewed through a microscope) are in exact alignment. The procedure is repeated with the gas of unknown density.

If P_1 and P_2 are the pressures of the two gases corresponding to the coincidence of E and F, then it follows that the two gases must have equal densities at these pressures.

For an ideal gas,
$$PV = nRT$$
$$= \frac{w}{M}RT$$

where w is the weight of gas present and M is the molar weight of the gas. Thus
$$P = \frac{w}{MV}RT = d\frac{RT}{M}$$

where d is the density of the gas.

For the two gases (at the same temperature and occupying the same volume) having equal densities at pressures P_1 and P_2 respectively,
$$d = \frac{M_1 P_1}{RT} = \frac{M_2 P_2}{RT}$$
or
$$\frac{P_1}{P_2} = \frac{M_2}{M_1}$$

A correction is applied in these experiments to allow for the deviation from ideal behaviour.

Example 21 At 0°C the exact alignment of the pointers of the microbalance occurred with oxygen at a pressure of 382·9 mm and with carbon monoxide at a pressure of 437·4 mm. To allow for the deviation from ideal behaviour, the ratio P_1/P_2 has to be reduced by a factor of 0·033%. Calculate the weight of one mole of carbon monoxide.

Solution
$$\frac{P_{\text{oxygen}}}{P_{\text{carbon monoxide}}} = \frac{382·9}{437·4} = 0·8754$$

$$\text{Correction factor} = \frac{0·8754 \times 0·033}{100} = 0·0003$$

Corrected ratio,
$$P_{\text{oxygen}} : P_{\text{carbon monoxide}} = 0·8751$$
Thus
$$\frac{M_{\text{carbon monoxide}}}{M_{\text{oxygen}}} = 0·8751$$

from which
$$M_{\text{carbon monoxide}} = 0\cdot8751 \times 31\cdot999$$
$$= 28\cdot00$$

EXERCISES

Take R as $0\cdot082$ l atm deg^{-1}.
1 mole of gas occupies $22\cdot4$ litres at s.t.p.

1. Find the final volume of $0\cdot1$ mole of oxygen, originally present at s.t.p., if:
   (a) the pressure is doubled,
   (b) the temperature is raised to 100°C,
   (c) the pressure is reduced to 50 cm and the temperature is lowered to $-93\cdot4$°C.

2. A flask of 3 litres capacity contains a mixture of oxygen and nitrogen in the proportion 1:2, and the pressure of the mixture is 15 lb in^{-2}. If this flask is connected to a second flask containing 2 litres of argon at a pressure of 10 lb in^{-2}, calculate the final pressure and the partial pressure of oxygen after mixing, assuming the temperature remains constant.

3. A Dumas bulb of $218\cdot5$ cm^3 capacity was sealed full of ethanol vapour at 98°C. The weight of the bulb and vapour was $52\cdot814$ g, while the weight of the bulb full of air was $52\cdot754$ g. If the laboratory temperature and pressure was 15°C and 750 mm, calculate the weight of one mole of ethanol vapour. (Density of air at s.t.p. is $0\cdot001293$ g cm^{-3}.)

4. How many moles of water are represented by the following?
   (a) $1\cdot00$ g of ice.
   (b) 200 cm^3 of steam at 127°C and $1\cdot64$ atm pressure.
   (c) 50 litres of oxygen saturated with water vapour at 18°C (vapour pressure of water at 18°C is $15\cdot4$ mm).

5. 25 cm^3 of acetic acid vapour diffused through a porous membrane in $16\cdot4$ minutes, while, under the same conditions, 25 cm^3 of methane took 6 minutes exactly. Show that these results are in accordance with the presence of double molecules of acetic acid $(CH_3COOH)_2$ in the vapour under these conditions.

6. The densities of methane and ammonia are $0\cdot717$ g l^{-1} and $0\cdot771$ g l^{-1} respectively at s.t.p. If the weight of one mole of methane is $16\cdot0$ g, what is the weight of one mole of ammonia?

7. A cylinder contains $1\cdot74$ kg of liquefied petroleum gas. Assuming

the gas to be pure butane, C_4H_{10}, calculate the volume this gas would occupy at one atmosphere pressure and 27°C.
8. Calculate the volume occupied by a mixture of 14 g nitrogen and 16 g oxygen at 8·2 atm and 127°C.
9. 0·305 g of an organic compound containing 12·8% carbon, 85·1% bromine and 2·1% hydrogen displaced 38·0 cm³ of air (corrected for the presence of water vapour) at a true pressure of 752 mm and 9°C in a Victor Meyer determination. Deduce the probable molecular formula of the compound.
10. A 200 cm³ vessel contains 1·00 g of argon and some acetone vapour. The vessel was cooled slowly until some liquid acetone was formed. If, at this point, the total pressure in the vessel was 239 cm and the temperature was 15°C, calculate the vapour pressure of acetone at 15°C. (Neglect the volume of liquid acetone.)
11. What is the most probable identity of a gaseous aliphatic hydrocarbon which diffuses at exactly the same rate as nitrogen?

If 25 cm³ of a mixture of this hydrocarbon and methane effuse through a fine orifice in 17·3 minutes, while the same volume of nitrogen effuses in 20 minutes under the same conditions, deduce the percentage composition of the hydrocarbon mixture.
12. Calculate the root mean square velocity of the molecules in methane at 127°C and at 76 cm pressure.
13. A flask weighed 124·75 g when full of air at 760 mm pressure, and 125·47 g when filled with carbon dioxide at the same pressure. When filled with air at exactly half an atmosphere pressure, the flask weighed 124·07 g. All the measurements were made at the same temperature. At this temperature and at one atmosphere pressure, air is 14·3 times as dense as hydrogen under the same conditions. Calculate the weight of one mole of carbon dioxide from these results.
14. Calculate the weight of one mole of toluene vapour from the following results obtained during a Dumas determination:

Weight of bulb full of air	25·230 g
Temperature of sealing	140°C
Weight of bulb and vapour	25·458 g
Weight of bulb full of water	179·00 g
Barometric pressure	758 mm
Laboratory temperature	18°C
Density of air at s.t.p.	0·001293 g cm^{-3}

(Density of water may be taken as 1·00.)
15. A closed vessel of 1 litre capacity is filled with nitrogen at s.t.p. Then 0·01 mole of liquid water is added and the vessel is sealed

and heated to 50°C. If the vapour pressure of water at 50°C is 92 mm, calculate:
(a) the number of moles of water in the vapour phase,
(b) the total pressure obtaining in the vessel at 50°C.
(Neglect any volume occupied by liquid water, and assume that water vapour behaves as an ideal gas.)

16. If the vapour pressures of water and bromobenzene at 95·3°C are 640 mm and 120 mm respectively, calculate the weight of bromobenzene carried over by 90 g of water in a steam distillation at 760 mm.

17. When 0·103 g of acetic acid was volatilized in a Victor Meyer apparatus, 21·5 cm^3 of air was displaced at a corrected barometric pressure of 748 mm and at 17°C. The vapour pressure of water is 15 mm at 17°C. Calculate the weight of one mole of acetic acid vapour and comment on the result.

18. How many moles of nitrogen must be added to a mixture of 4 g of hydrogen and 14 g of carbon monoxide contained in a 50-litre flask at 27°C to bring the total pressure up to 1·5 atm?

19. A 10 cm^3 flask contains water vapour, carbon dioxide and nitrogen at a pressure of 25·3 mm at 27°C. The contents of the flask were transferred to a cold trap at $-10°C$ where all the water vapour froze out. When the gas mixture was returned to the original flask, the pressure measured 21·3 mm at 27°C. The carbon dioxide was then removed by freezing in a similar way at $-112°C$, and the remaining nitrogen exerted a pressure of 12·5 mm in the original flask at 27°C. Calculate the total number of moles of gas present and the mole fraction of each gas.

20. 5 g of pure calcium carbonate was totally decomposed by excess hydrochloric acid in a closed 5-litre vessel at 9°C. If the pressure of air in the vessel before reaction was 733 mm, calculate the final pressure in the vessel and the partial pressure of the carbon dioxide produced by the reaction. (Neglect the volume of the liquid phase, and assume that carbon dioxide does not dissolve in the liquid phase.)

21. Using a gas density microbalance, the densities of carbon dioxide and ethene are found to be equal at pressures of 272·6 mm and 427·5 mm respectively. A correction factor, which reduces the ratio $P_{CO_2}:P_{C_2H_4}$ by 0·0006 is applied. Calculate the weight of one mole of ethene, given that one mole of carbon dioxide weighs 44·01 g.

3. Liquids and Solutions

The liquid state is characterized by the following properties:

1. The molecules in a liquid are packed much more closely than in a gas, and thus the density of a liquid is much greater than that of a gas (at normal pressures). The volume of a liquid does not alter significantly with increasing pressure.
2. The molecules in a liquid are in a state of random motion, although this is much more restricted than in the gaseous state. Hence, liquids tend to diffuse into each other on mixing.
3. The unbalanced forces of attraction at the edges of a liquid (Figure 13) are more pronounced than in a gas, and so liquids have sharp boundaries and exhibit the phenomenon of surface tension.
4. Liquids have no rigidity or elasticity—they flow under the influence of a force.

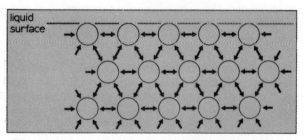

Fig. 13. Unbalanced forces of attraction at a liquid surface

The Vapour Pressure of a Liquid

Since the molecules in a liquid are in constant motion, some molecules can acquire sufficient energy to break free of the unbalanced forces of attraction at the surface of the liquid and escape into the gaseous phase. In a closed system, some molecules in the gaseous phase return to the liquid phase, and eventually a dynamic equilibrium is set up in which the number of molecules escaping from the liquid is equal to the number returning (Figure 14). At this stage, the vapour is said to be *saturated*, and the pressure exerted by the liquid molecules present in the vapour phase is called the *saturated vapour pressure* of the liquid at that temperature. Provided some liquid is present, the saturated vapour pressure is independent of the amount of liquid present.

An increase in molecular energy corresponds to a rise in temperature, and this results in a greater number of molecules having enough energy to break through into the gaseous phase. Consequently, the saturated vapour pressure rises as the temperature is increased. When the saturated vapour pressure is equal to the pressure standing above

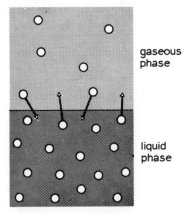

Fig. 14. Dynamic equilibrium set up by molecules traversing a gas–liquid boundary at equal rates

the liquid (i.e. the external or atmospheric pressure) bubbles of vapour form within the liquid, and the liquid is said to boil. The *boiling point* of a liquid is defined as *the temperature at which its saturated vapour pressure is equal to the external pressure*. A graph showing the variation of vapour pressure with temperature is useful in estimating the boiling point of a liquid at various pressures, as illustrated in the following example.

Example 22 Express the following vapour pressure–temperature data for chloroform in the form of a graph, and deduce (a) the boiling point of this liquid at 76 cm pressure, and (b) the pressure at which the boiling point of chloroform is 40°C.

Temperature (°C)	9·5	25·0	35·0	42·3	48·4	53·8	58·3	62·1
Vapour pressure (*cm mercury*)	10	20	30	40	50	60	70	80

Solution   The graph of the vapour pressure data is shown in Figure 15. From this graph, we deduce the following:

(a) The temperature at which the vapour pressure of the liquid reaches 76 cm is 61°C, which is the boiling point at this pressure.

(b) At 40°C, chloroform has a vapour pressure of 36·5 cm. For chloroform to boil at 40°C, the external pressure must be reduced to 36·5 cm.

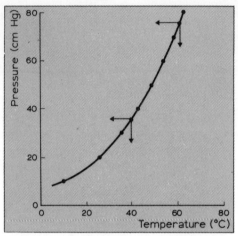

Fig. 15. Vapour pressure–temperature curve for chloroform

Solutions

A solution is regarded as a completely uniform mixture of two or more substances. The term uniform is used to express the fact that it is not possible to discern any inhomogeneity in properties such as density, refractive index, concentration or appearance in different samples drawn from various parts of the solution. At the molecular level, a solution cannot be perfectly homogeneous.

The terms *solute* and *solvent* are used in connection with solutions, and it is the solute which is dissolved by the solvent, the latter being the substance which is present in excess.

The concentration of a solution is the amount of solute present in a given quantity of solution, and this quantity can be expressed in several ways:

1. *Percentage composition:*
   (a) by weight (w/w). This is the number of grammes of solute contained in 100 g of solution.
   (b) by volume (v/v). This is the volume of solute present in 100 volumes of solution.
   (c) as a weight/volume ratio (w/v). This is the number of grammes of solute present in 100 cm^3 of solution.
2. *Grammes per litre*, i.e. the number of grammes of solute per litre of solution.
3. *Moles per litre*, i.e. the number of moles of solute per litre of solution. This quantity expresses the molarity of a solution. The notation [X] is used to represent the concentration of a species X in moles per litre of solution. Hence, in 0·1M hydrochloric acid, $[H^+] = 0·1$.

   Molality: this is the number of moles of solute dissolved in 1000 g of solvent.
5. *Normalities.* A normal solution contains one g-equivalent per litre of solution. However, a substance may have more than one equivalent weight depending on the reaction it undergoes; for example, formic acid has an equivalent weight of 46 when reacting as an acid and 23 when reacting as a reducing agent. Consequently, the use of normalities to denote the concentration of a solution in general terms can be ambiguous.

Types of solution   There are five general types of solution:

1. gases in gases
2. gases in liquids
3. liquids in liquids
4. solids in liquids
5. solids in solids

Solutions of Gases in Gases

All gases mix, and the relative proportions of each component may be expressed in terms of a partial pressure or mole fraction. Molecules of a liquid may escape into the gaseous phase at temperatures below the boiling point of the liquid (see p. 54) producing a vapour. The saturated vapour pressure of a liquid at different temperatures may be measured by a transpiration method, or by using an isoteniscope.

Transpiration method   A measured volume of dry air (at a known pressure) is slowly bubbled through a series of bulbs containing water at a controlled temperature. In this way, the air becomes saturated with water vapour. The saturated air is now passed through a series of calcium chloride tubes where the water vapour in the gas stream is absorbed. From the weight of water vapour taken up by the measured volume of dry air, the vapour pressure of water at the temperature of the experiment may be calculated.

Example 23 15·84 litres of dry air (measured at s.t.p.) were bubbled through water at 25°C and then through a system of calcium chloride tubes. A gain in weight of 0·4205 g in the latter was noted. Calculate the saturated vapour pressure of water at 25°C.

Solution

$$\frac{p_{H_2O}}{P} = \frac{n_{H_2O}}{n_{H_2O} + n_{air}}$$

$$\text{Number of moles of water} = \frac{0 \cdot 4205}{18} = 0 \cdot 0233$$

$$\text{Number of moles of air} = \frac{15 \cdot 84}{22 \cdot 4} = 0 \cdot 707$$

The total pressure is 760 mm. Hence

$$p_{H_2O} = \frac{0 \cdot 0233 \times 760}{(0 \cdot 0233 + 0 \cdot 707)} = 24 \cdot 3 \text{ mm}$$

The isoteniscope A common form of the isoteniscope is shown diagrammatically in Figure 16. The bulb is half filled with the liquid under test, which is then heated to boiling several times in order to displace air from the bulb and U-tube. On cooling, some liquid condenses in the U-tube. By tilting the apparatus, liquid can be transferred from the bulb to the U-tube (and vice versa) and in this way the liquid level in the U-tube is adjusted to that shown. It is essential that the vapour space between the bulb and the right-hand limb of the U-tube is completely free from air. The apparatus is placed in a thermostat and a partial vacuum is applied. The pressure on the left-hand side of the system is adjusted by means of the three-way tap until the liquid levels are equal. At this point, the air pressure shown by the manometer is equal to the vapour pressure in the vapour space above the liquid in the bulb. The cold trap prevents the escape of vapour into the manometric and vacuum systems.

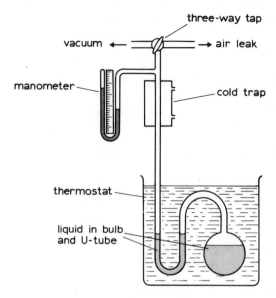

Fig. 16. Isoteniscope

Solutions of Gases in Liquids

The solubility of a gas in a liquid depends on the nature of the gas and solvent, and on the temperature and pressure. With water as a solvent, helium is the least soluble and ammonia the most soluble gas.

Usually, heat is liberated when a gas dissolves in water and in accordance with Le Chatelier's principle (p. 153) a decrease in solubility of the gas would be expected following a rise in temperature. This is found to be generally true, although the solubility of hydrogen and the noble gases decreases at first, then increases as the temperature rises. When a beaker of tap water is heated, the dissolved air is expelled as the water warms, and bubbles of air collect on the base and sides of the beaker; hence the use of boiled-out water when making up certain solutions such as potassium permanganate.

From kinetic considerations, it would be expected that an increase in pressure should increase the solubility of a gas, as on compression the rate of gas molecules striking the liquid interface is increased and the extent of absorption of these molecules into the liquid phase is greater. The effect of pressure on gas solubility is described by Henry's law.

Henry's law *The weight of a gas dissolved by unit volume of liquid is directly proportional to the pressure of the gas with which it is in equilibrium, the temperature being constant.*

Thus, the mass of gas dissolved is proportional to pressure, or

$$m = kp$$

The solubility of a gas in a liquid is usually given in terms of an *absorption coefficient* (α), which is the maximum volume of gas (reduced to s.t.p.) which can be dissolved by 1 cm³ of liquid at a particular temperature.

Determination of the solubility of sparingly soluble gases The apparatus used for this determination is shown in Figure 17, and the experiment is carried out in the following stages.

1. The volume of the absorption vessel T is determined by filling with water and either measuring the volume used or by weighing before and after filling.
2. The absorption vessel is filled with boiled-out de-ionized water, and the apparatus is placed in an air thermostat to attain the desired temperature.
3. R is filled completely with mercury by raising the tube S. Tap A is connected to a flask of gas under test, and, by lowering S, the gas is drawn into R.

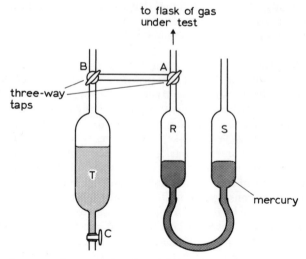

Fig. 17. Apparatus for measuring the solubility of a sparingly soluble gas

4. Tap A is closed and the mercury levels in R and S are adjusted so that the volume of gas at atmospheric pressure in R can be read off.
5. S is raised to provide a head of mercury and the taps A and B are turned to connect R and T. At the same time, tap C is opened and a volume of water is run out of T. In this way a volume of gas is transferred from R to T. All the taps are now closed.
6. The absorption vessel is now shaken for some time, care being taken to ensure that heat from the hands does not change the temperature of T and that the liquid does not splash into the tube below B.
7. R and T are re-connected and the volume in R is read off (the mercury levels in R and S being equal). The gas is transferred back to T and re-shaken.
8. When no further diminution in volume is noticed, the water in T is saturated with gas, and the solubility is calculated as shown in Example 24.

Example 24 The following results were obtained using the above method to determine the solubility of oxygen in water at 20°C.

Volume of absorption vessel T	53·35 cm^3
Volume of water displaced	21·40 cm^3
Initial volume of oxygen in R	32·64 cm^3
Final volume of oxygen in R after absorption	10·15 cm^3
Atmospheric pressure	742 mm mercury

Solution The volume of water remaining in the absorption vessel in which the oxygen dissolves is

$$53\cdot35 - 21\cdot40 = 31\cdot95 \text{ cm}^3$$

Volume of oxygen originally present was 32·64 cm^3, and the volume of oxygen remaining after absorption is

$$21\cdot40 + 10\cdot15 = 31\cdot55 \text{ cm}^3$$

Thus 31·95 cm^3 of water dissolve

$$32\cdot64 - 31\cdot55 = 1\cdot09 \text{ cm}^3 \text{ of oxygen}$$

and 1 cm^3 of water would dissolve

$$\frac{1\cdot09 \times 742 \times 273}{31\cdot95 \times 760 \times 293}$$

$$= 0\cdot031 \text{ cm}^3 \text{ of oxygen at s.t.p.}$$

Hence, the absorption coefficient of oxygen at 20°C is 0·031.

Determination of the solubility of highly soluble gases This is done by measuring the concentration of a saturated solution of the gas in the solvent. The gas is bubbled through the solvent until a saturated solution is produced, and the concentration of this solution is found by analytical techniques including volumetric or gravimetric methods. Alternatively, a physical property such as refractive index or density may be measured and the result compared with a previously obtained calibration chart.

Example 25 9·50 g of a solution saturated with ammonia gas at s.t.p. was made up to one litre by adding de-ionized water. 25·0 cm^3 of this diluted solution titrated against 32·35 cm^3 of M/5 hydrochloric acid. Calculate the solubility of ammonia in water, expressing the result in terms of

  (a) grammes of ammonia dissolved by 100 g of water,
  (b) an absorption coefficient.

Solution (a) If 25·0 cm^3 of dilute solution titrated against 32·35 cm^3 M/5 hydrochloric acid then 1000 cm^3 of ammonia solution would titrate against

$$\frac{32\cdot35 \times 1000}{25} \text{ cm}^3 \text{ M/5 acid}$$

$$= \frac{32\cdot35 \times 1000}{25 \times 1000} \times \frac{1}{5} \text{ mole HCl} = 0\cdot259 \text{ mole HCl}$$

But 1 mole HCl reacts with 1 mole NH$_3$, from the equation

$$NH_3 + HCl = NH_4Cl$$

so that 1000 cm^3 dilute solution contains 0·259 mole NH$_3$, which is

$$0\cdot259 \times 17 = 4\cdot40 \text{ g NH}_3$$

If 9·50 g of saturated solution contains 4·40 g of ammonia, then 9·50 − 4·40 or 5·1 g water dissolve 4·40 g of ammonia, and 100 g of water would dissolve

$$\frac{4\cdot40 \times 100}{5\cdot1} = 86\cdot3 \text{ g of ammonia}$$

 (b) If 5·1 g (i.e. 5·1 cm^3) of water dissolved 4·40 g of ammonia, then 1·0 cm^3 of water would dissolve

$$\frac{4\cdot40 \times 1 \times 22\,414}{5\cdot1 \times 17} = 1140 \text{ cm}^3 \text{ of ammonia at s.t.p.}$$

Thus the absorption coefficient of ammonia in water at 0°C and one atmosphere pressure is 1140.

Solutions of mixtures of non-reacting gases Dalton's extension to Henry's law states: *for a mixture of non-reacting gases (assumed ideal) in equilibrium with a liquid, the mass of any one gas dissolved from the mixture is proportional to the partial pressure of that gas.*

Both this statement and Henry's law apply only to ideal gases. Deviations from these laws are to be expected at high pressures and low temperatures and with gases that are easily liquefied (i.e. have strong intermolecular forces of attraction). Sparingly soluble gases obey the laws satisfactorily at normal temperatures and pressures.

Example 26 At 20°C the absorption coefficients of nitrogen, oxygen and carbon dioxide are 0·016, 0·031 and 0·878 respectively. If a gas mixture containing 79% nitrogen, 20·5% oxygen and 0·5% carbon dioxide is shaken with 100 cm^3 of water at 20°C and 5 atm pressure, calculate the volumes (corrected to s.t.p.) of each gas that would dissolve. Find also the composition of the gas expelled on heating the water to boiling point, neglecting the vapour pressure of water.

Solution If the vapour pressure of water is ignored, the total pressure of the gas mixture is 5 atm.

$$\text{Partial pressure of nitrogen} = \frac{79 \times 5}{100} = 3\cdot95 \text{ atm}$$

$$\text{Partial pressure of oxygen} = \frac{20\cdot5 \times 5}{100} = 1\cdot025 \text{ atm}$$

$$\text{Partial pressure of carbon dioxide} = \frac{0\cdot5 \times 5}{100} = 0\cdot025 \text{ atm}$$

When reduced to s.t.p.,

volume of nitrogen dissolving = $3\cdot95 \times 100 \times 0\cdot016 = 6\cdot32$ cm^3

Similarly,

volume of oxygen dissolving = $1\cdot025 \times 100 \times 0\cdot031 = 3\cdot18$ cm^3

and

volume of carbon dioxide dissolving = $0\cdot025 \times 100 \times 0\cdot878$
 = $2\cdot19$ cm^3

The total volume of gas dissolved is 11·69 cm^3, and this has the composition

$$\text{nitrogen} \quad \frac{6\cdot32 \times 100}{11\cdot69} = 54\cdot1\%$$

oxygen $\quad\dfrac{3\cdot18\times100}{11\cdot69}=27\cdot2\%$

carbon dioxide $\quad\dfrac{2\cdot19\times100}{11\cdot69}=18\cdot7\%$

and this would be the composition of the gas expelled on boiling.

Solubility of gases in solutions The solubility of a gas is usually much lower in a salt solution than in pure water. This is known as the *salting out effect* and, from a study of a range of salt solutions, it appears that the salting out effect is greater in the presence of ions of (a) high charge and (b) small size. If the solution of a gas is effected by a solvation (see p. 119) of the gas molecules by the polar solvent molecules, the addition of ions, which are preferentially solvated, will reduce the degree of solvation of the gas molecules, and thus lower the mass of gas held in solution. A useful feature of the salting out effect is that it allows moderately soluble gases such as chlorine or carbon dioxide to be collected over saturated sodium chloride solution without too much loss of gas—an important factor when samples are withdrawn from a gas stream for analysis.

Solutions of Liquids in Liquids

Liquids which mix with each other completely over all ranges of composition are said to be fully miscible. If one liquid is partially or totally insoluble in the other, the terms partially miscible and immiscible respectively are used.

The vapour pressure of a solution of two miscible liquids In a mixture of two miscible liquids (say A and B) the ability of molecules of A to escape from the liquid into the gaseous phase is impaired by the presence of B, and vice versa. For an ideal solution, that is one in which no intermolecular forces exist, the vapour pressure of each constituent is given by Raoult's law:

Raoult's law *The vapour pressure of a constituent in an ideal solution is equal to the vapour pressure exerted by the pure constituent at that temperature, multiplied by the mole fraction by which it is present.*

Stated mathematically,

$$p_A = x_A P_A^0$$

where p_A is the vapour pressure of A over the liquid mixture
$\quad\quad x_A$ is the mole fraction of A in the liquid mixture
$\quad\quad P_A^0$ is the vapour pressure of pure A at that temperature.

No liquid mixture is ideal, and deviations from Raoult's law are observed, but the law holds well for dilute solutions.

Example 27 Calculate the vapour pressure over a solution containing 11·7 g benzene and 4·6 g toluene at 50°C, if the vapour pressure of the pure components at this temperature are 272 mm and 84 mm respectively.

Solution

$$\text{Number of moles of benzene present} = \frac{11\cdot7}{78} = 0\cdot15$$

$$\text{Number of moles of toluene present} = \frac{4\cdot6}{92} = 0\cdot05$$

Hence,

Total number of moles = 0·20

$$\text{Mole fraction of benzene } (x_A) = \frac{0\cdot15}{0\cdot20} = 0\cdot75$$

$$\text{Mole fraction of toluene } (x_B) = \frac{0\cdot05}{0\cdot20} = 0\cdot25$$

Partial vapour pressure of benzene, $p_A = x_A P_A^0 = 0\cdot75 \times 272$
$$= 204 \text{ mm}$$

Similarly,

Partial vapour pressure of toluene = $0\cdot25 \times 84 = 21$ mm

Thus

Total vapour pressure of the mixture = 225 mm

Similar calculations may be made over the complete range of compositions, and the results (which are of course for an ideal case) are expressed graphically as shown in Figure 18. The straight line B represents the increase in the partial vapour pressure of benzene as the mole fraction of benzene in the mixture steadily increases. T is the corresponding line showing the steady decrease in the partial vapour pressure of toluene over the mixture as the mole fraction of toluene in the mixture is lowered.

Boiling point–liquid composition curves It is clear from Figure 18, that at 50°C pure benzene has a vapour pressure greater than any of the benzene–toluene mixtures, and as the temperature rises pure benzene will continue to show the greatest vapour pressure. Consequently, pure benzene will have a vapour pressure of 760 mm (or will equal the external pressure) at a lower temperature than any of the mixtures of benzene and toluene. This means that pure benzene has the lowest boiling point and pure toluene the highest boiling point of all the various compositions. The boiling points of the various liquid mixtures lie between these two extreme values; mixtures rich in benzene will boil at a lower temperature, while those rich in toluene will have a higher boiling point. This is interpreted by means of a boiling point–liquid composition graph (which is the inverse of the graph of vapour pressures shown in Figure 18), as shown in Figure 19. From this graph, a mixture containing 0·25

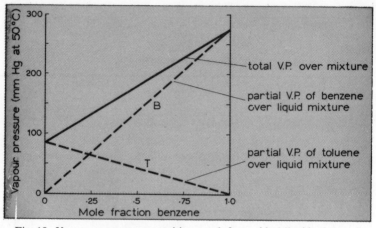

Fig. 18. Vapour pressure–composition graph for an ideal liquid mixture at 50 °C

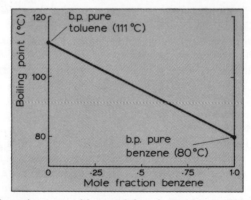

Fig. 19. Boiling point–composition graph for mixtures (assumed ideal) of benzene and toluene

mole of benzene and 0·75 mole of toluene boils at 103°C, but it should be noted that the graph is constructed on the assumption that Raoult's law is obeyed over the whole range of compositions.

Composition of the vapour formed by boiling an ideal liquid mixture
Suppose a liquid mixture contains an equal number of moles of two components A and B, and let the vapour pressure of pure A be 100 cm, and that of pure B be 52 cm at a temperature of 120°C. Then the partial pressure of A in the vapour is

$$\frac{50 \times 100}{100} = 50 \text{ cm}$$

and that of B is

$$\frac{50 \times 52}{100} = 26 \text{ cm}$$

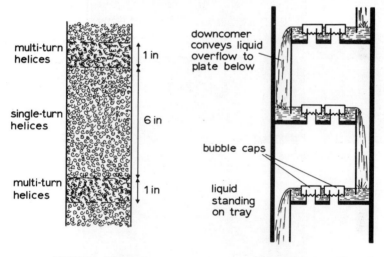

PACKED COLUMN   BUBBLE CAP COLUMN
Fig. 20. Fractionation columns

This makes the total pressure 76 cm, and this particular mixture boils at 120°C under one atmosphere pressure. The vapour produced will not have the same composition as the boiling liquid, but will contain amounts of A and B in proportion to their partial vapour pressures; i.e.

$$\text{mole fraction of A in vapour} = \frac{50}{76} = 0 \cdot 66 \text{ or } 66 \text{ mole \%}$$

and

mole fraction of B in vapour = $\dfrac{26}{76}$ = 0·34 or 34 mole %

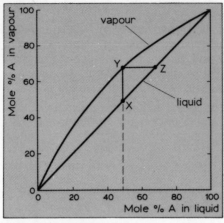

(a) Liquid–vapour composition curves

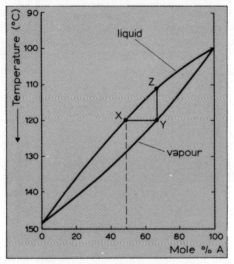

(b) Temperature–composition curves

Fig. 21

The boiling liquid contains 50 mole % of A, but the vapour produced contains 66 mole % of A. This enrichment on vaporization of the low boiling component forms the basis of fractional distillation, and this degree of enrichment (produced by boiling a liquid and condensing the vapour in equilibrium with it) is known as that produced by a *theoretical plate*.

If the enriched vapour is condensed and reboiled, a further degree of enrichment is produced, and a device for accomplishing this is a fractionation column. Typical columns may be packed with glass beads, glass or metal rings, helices, etc. in order to increase the area of liquid in contact with the vapour. Larger columns contain bubble cap trays in which the vapour from a boiling liquid is made to bubble through the liquid standing on the bubble cap tray. Figure 20 shows diagrams of both types of column. In the column packed with helices the multi-turn helices assist column drainage and help to prevent the formation of liquid locks.

The progress of fractionation can be checked by means of a liquid–vapour composition curve as shown in Figure 21. The boiling liquid has the composition shown by the point X (50 mole % of A) and this produces a vapour of composition Y. This is condensed to give a liquid of the same composition which is represented by point Z. The complete step XYZ is the enrichment produced by a theoretical plate. The enrichment produced by an actual bubble cap plate is less than theoretical; for example, a column containing 20 actual plates may give a separation equal to 14 theoretical plates. For a packed column, H.E.T.P (height equivalent per theoretical plate) values are quoted; for instance, a laboratory column 30 inches in packed length may produce an enrichment corresponding to 5 theoretical plates, and thus it has an H.E.T.P value of 6 inches per plate.

Derivation of Raoult's law If the vapour standing over a liquid behaves as an ideal gas, then the pressure of the vapour is exactly proportional to the number of molecules per unit volume in the vapour phase. In turn, the number of molecules entering the vapour phase is proportional to *that fraction* of molecules in the liquid which can acquire sufficient energy to escape from the liquid phase.

If the liquid phase is made up of n_A molecules of solvent and n_B molecules of a *non-volatile* solute, then the fraction of molecules in solution which have the ability to escape is

$$\frac{n_A}{n_A + n_B}$$

so that

$$p_A \propto \frac{n_A}{n_A + n_B}$$

or
$$p_A = k\frac{n_A}{n_A + n_B} = kx_A$$
where x_A is the mole fraction of A.

When the pure solvent alone is present, $x_A = 1$ and $p_A = P_A^0$ so that $P_A^0 = k$ and
$$p_A/P_A^0 = x_A$$
which is Raoult's law.

(The same argument applies to cases in which the solute is volatile, the total vapour pressure over the liquid mixture then being $p = p_A + p_B$.)

In this derivation, it has been assumed that the ability of a molecule of the solvent to escape remains unaltered on the addition of the second component. (This is implied by assuming the proportionality constant k to be a true constant.) The assumption is not justified, and this leads to the observed deviations from Raoult's law.

Deviations from Raoult's law Some liquid mixtures give total vapour pressures which are higher than or less than the values predicted by Raoult's law. These high and low values (described as positive or negative deviations respectively) may lead to the formation of a maximum or minimum vapour pressure for the mixture at a certain composition, as shown in Figure 22.

These deviations are due to a change in the ability of the molecules of one component (A) to escape from the liquid phase following the addition of the second component (B). For example, if the solvent

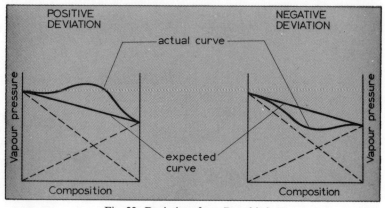

Fig. 22. Deviations from Raoult's law

(component A) is polar the ability of molecules to escape from this environment is impaired by the fact that they have to escape against strong attractive forces. On addition of a non-polar solute, B, the extent of these attractive forces will be lessened by dilution, and more molecules of A will escape into the vapour phase leading to a positive deviation from Raoult's law.

Positive deviations To give a positive deviation from Raoult's law, one of the components is normally of the type in which the molecules have to escape into the vapour phase against forces which are larger than usual. Liquids

   (a) in which strong intermolecular forces of attraction obtain (e.g. polar substances, liquids in which the molecules are associated), or
   (b) which contain long chain molecules,

are typical examples. On mixing these liquids with a second component which (a) is not polar, or (b) contains small molecules, a system is formed in which the vapour pressure is in excess of that expected, since mixing tends to ease the restraint suffered by the escaping molecules in the first component. For example, n-heptane and n-hexane behave ideally, but n-heptane and ethanol mixtures show a large positive deviation from Raoult's law. In the second system, ethanol is more polar than n-heptane while the chain lengths of the molecules are quite different. Systems showing positive deviations are quite common.

Negative deviations If the two components A and B are such that

   (a) there is a strong intermolecular attraction between A and B, or
   (b) there is the possibility of some type of compound formation between A and B in the liquid phase,

then the vapour pressure of the solution will be less than that calculated from Raoult's law, since mixing A and B makes it more difficult for the molecules to escape from the liquid phase. Systems which show this effect may contain an acidic and a basic component (e.g. pyridine and acetic acid) or a substance which forms a hydrate with water (e.g. hydrochloric acid and water).

In many cases, factors operate to produce both positive and negative deviations from Raoult's law. Occasionally these balance out, giving a pseudo-ideal mixture. The extents of both positive and negative deviations decrease as the dilution of the solution increases, and Raoult's law is obeyed in dilute solutions.

Examples of deviations from Raoult's law A negative deviation from Raoult's law leads to a mixture showing a maximum boiling point—an example of this type of behaviour is shown by the nitric acid and water system. The data for this system are:

Mole fraction HNO_3 in vapour	Mole fraction HNO_3 in liquid	Temp. °C
0·000	0·000	100
0·006	0·084	106
0·020	0·120	112
0·160	0·300	121·5
0·380	0·380	122
0·600	0·400	121
0·760	0·460	118
0·890	0·530	112
0·920	0·610	99
0·960	0·930	90
1·000	1·000	86

The pressure is constant at 760 mm.

These values are plotted on the graph shown in Figure 23.

When a liquid mixture containing less than 38 mole % of nitric acid is distilled, the vapour produced is richer in water than the boiling liquid. Subsequent fractionation of the condensed vapour will produce pure water. The residual liquid becomes gradually richer in nitric acid and the boiling point rises steadily to 122°C. At this temperature, the liquid contains 38 mole % of nitric acid, and the graph shows that the vapour produced at this composition by

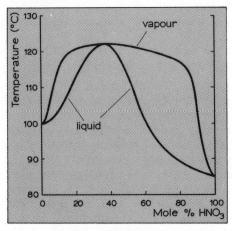

Fig. 23. Maximum boiling point mixture: nitric acid and water

the boiling liquid has exactly the same composition as the liquid. This maximum boiling point mixture is referred to as an *azeotrope*. Mixtures containing more than 38 mole % of nitric acid boil at temperatures below 122°C, and again, the liquid composition tends to reach the azeotropic composition on boiling, while the vapour can be fractionated to give pure nitric acid. It is a feature of all mixtures which form an azeotrope, that they cannot be separated into two pure components by fractionation.

Mixtures of hydrochloric acid and water, or chloroform and water are other examples of systems which form a constant boiling or azeotropic mixture with a maximum boiling point.

A positive deviation from Raoult's law (Figure 22) leads (if the deviation is large enough) to the mixture showing a minimum boiling point. The carbon disulphide–acetone system is a typical example:

Mole fraction CS_2 in vapour	Mole fraction CS_2 in liquid	Temp. °C
0·00	0·00	56
0·18	0·05	51
0·35	0·13	46
0·44	0·18	44
0·53	0·29	41
0·57	0·38	40
0·60	0·45	39·5
0·66	0·66	39
0·76	0·88	40·5
0·88	0·97	43
1·00	1·00	46

The pressure is constant at 760 mm.

These values are shown graphically in Figure 24. The azeotropic mixture in this case contains 66 mole % of carbon disulphide and boils at the minimum temperature of 39°C. Again, fractionation of a mixture of composition other than 66 mole % of carbon disulphide eventually produces a liquid containing either pure carbon disulphide or acetone (depending on which side of the azeotropic composition the initial liquid composition lies) and a vapour of the azeotropic composition.

Many other examples of this type are known, for example ethanol and water, tetrachloromethane and ethyl acetate, and n-propanol and water, all form constant boiling mixtures with a minimum boiling point.

Separation of the components of an azeotropic mixture An azeotrope may be distinguished from a pure compound since the composition

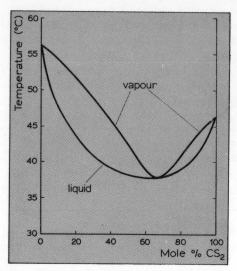

Fig. 24. Minimum boiling point mixture: carbon disulphide and acetone

of the constant boiling mixture changes if the pressure is either increased or decreased. However, it is not possible to separate the two components of a binary azeotrope by fractionation, so that the following alternative methods have to be employed.

(a) *Extractive distillation with a third component.* For example, the ethanol–water system forms an azeotrope containing 95·5% of alcohol, at 760 mm, boiling at 78·1°C. A small quantity of benzene is added to this mixture, and three separate fractions containing (i) an ethanol, benzene and water three-component azeotrope, (ii) an ethanol and benzene two-component azeotrope, and (iii) pure ethanol, can be distilled off at successively increasing temperatures.
(b) *Extraction* of one component of the azeotrope using an immiscible solvent.
(c) *Adsorption* of one component, using silica gel, for example.
(d) *Chemical methods.* For example, the water in the ethanol–water azeotrope can be removed by adding a suitable solid desiccant (e.g. calcium oxide).

Partially miscible liquids Pairs of liquids which dissolve in each other to a limited extent are termed partially miscible. For example, if a little phenol is added to water, the phenol dissolves completely.

If a larger quantity of phenol is used, two liquid layers are produced, one being a saturated solution of phenol in water, the other a saturated solution of water in phenol (Figure 25). These solutions are sometimes known as *conjugate solutions*.

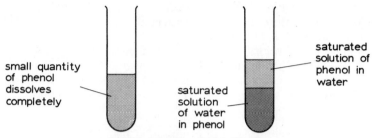

Fig. 25. Partially miscible liquids

The effect of temperature changes on conjugate solutions Generally, as the temperature is raised, the mutual solubility of the liquids increases, and a graph of the compositions of the liquid layers at various temperatures takes the form of curve shown in Figure 26, which represents the system formed by o-cresol and water.

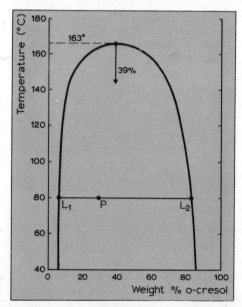

Fig. 26. Variation of partial miscibility with temperature; the o-cresol–water system

As the temperature rises, the two liquid layers become increasingly soluble—the water layer contains a greater proportion of o-cresol, while the organic layer shows an increasing water content. At 163°C, both layers contain 39% by weight of o-cresol, and the two liquid layers mix completely. This temperature is known as the *critical solution temperature* (C.S.T.) or *consolute temperature*, and it is the highest temperature at which a system of two partially miscible liquids can exist as two layers.

Liquid compositions at temperatures represented by points outside the curve correspond to one liquid layer only, while a system represented by a point inside the curve, say P, separates out into two liquid layers, of composition L_1 and L_2. It can be shown that

$$\frac{\text{weight of layer of composition } L_1}{\text{weight of layer of composition } L_2} = \frac{PL_2}{PL_1}$$

To obtain the data from which Figure 26 was plotted, the two liquids are shaken together at a controlled temperature until equilibrium has been established. The two layers are then separated and analysed. Alternatively, the temperature at which the two layers just begin to form (shown by the appearance of a turbidity on cooling the clear solution) is determined for a range of mixtures. In both methods, the experiments are repeated to give a large number of points on the graph.

In some instances, two partially miscible liquids become more miscible on cooling; for example the system 2:4:6 trimethyl pyridine and water shows a *lower critical solution temperature*, as shown in Figure 27. Below 5·7°C all mixtures of these two liquids are completely miscible.

A few liquid mixtures exhibit both an upper and a lower critical solution temperature; this type of behaviour is shown by nicotine and water, and by glycerol and guaiacol, the curve for which system is shown in Figure 28. A mixture of 37% by weight of glycerol forms one layer below 40°C and above 82°C, but produces two layers between these temperatures.

Solutions of Solids in Liquids

A solid which dissolves in a liquid is said to be soluble, while one which does not is termed insoluble. However, minute amounts of a substance normally regarded as insoluble do dissolve, while there is a limit to the amount of a soluble solid that can dissolve—this maximum quantity dissolved produces a *saturated solution*. Generally, more solid dissolves on warming a solution, and the extra amount crystallizes out when a saturated solution is cooled. In some

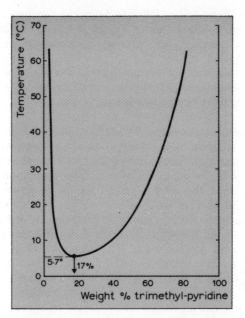

Fig. 27. Lower C.S.T. 2:4:6 trimethyl pyridine–water system

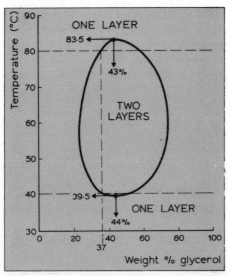

Fig. 28. Glycerol–guaiacol; a system showing upper and lower C.S.T.s

instances, such crystallization fails to occur spontaneously, and a *super-saturated* solution results. Vigorous stirring or seeding with a small crystal causes the excess solute to precipitate from a supersaturated solution. In view of this, the *solubility* of a solid is defined as *the maximum weight of solute which will dissolve in 100 g of solvent, at a given temperature, to produce a saturated solution, which is in contact with excess undissolved solute.*

Experimental determination of the solubility of a solid Most methods involve the analysis of a saturated solution, using volumetric or gravimetric methods. Since excess solid material has to be present in accordance with the definition of solubility, precautions have to be taken to prevent this excess solid from being removed with the solution withdrawn for analysis.

A saturated solution containing an excess of the undissolved solid is prepared at the temperature required. Some time is allowed for the system to reach equilibrium, after which the solution is decanted through a filter or a sample is withdrawn by means of a pipette (on the end of which is fitted a 'guard' tube containing a glass wool filter). The guard tube is removed before the contents of the pipette are discharged, and all apparatus used must be at the same temperature as the solution. The sample is then weighed, and analysed by volumetric or gravimetric methods, or evaporated very carefully to dryness. From these results, the solubility of the substance is found.

Example 28 25·2 g of a solution saturated with copper (II) sulphate, at 35°C, was made up to 200 cm^3 with de-ionized water. 25·0 cm^3 of the diluted solution reacted with potassium iodide solution to liberate iodine which titrated against 33·5 cm^3 of 0·118M sodium thiosulphate solution. Calculate the solubility of copper (II) sulphate at 35°C.

Solution If 25·0 cm^3 dilute solution titrated against 33·5 cm^3 0·118M thiosulphate, then 200 cm^3 dilute solution would titrate against

$$\frac{33 \cdot 5 \times 200}{25} \times \frac{0 \cdot 118}{1000} = 0 \cdot 0316 \text{ mole thiosulphate}$$

The equations for the analytical reactions are:

$$2CuSO_4 + 4KI = 2K_2SO_4 + 2CuI + I_2$$

and

$$I_2 + 2Na_2S_2O_3 = 2NaI + Na_2S_4O_6$$

Thus, 1 mole of copper sulphate liberates iodine which titrates against 1 mole of sodium thiosulphate. Hence, 0·0316 mole of copper (II) sulphate are present in 200 cm³ of dilute solution, or in 25·2 g of saturated solution. Since 1 mole of copper (II) sulphate weighs

$$63·5 + 32 + 64 = 159·5 \text{ g}$$

then 25·2 g saturated solution contains

$$159·5 \times 0·0316 = 5·04 \text{ g copper (II) sulphate}$$

Hence, 25·2 − 5·04 or 20·16 g water dissolve 5·04 g copper sulphate, so that 100 g water would dissolve

$$\frac{100 \times 5·04}{20·16} = 25·0 \text{ g copper (II) sulphate}$$

Thus the solubility of copper (II) sulphate at 35°C is 25·0 g/100 gH₂O.

For sparingly soluble salts other methods (see p. 201) have to be used.

Solubility curves The variation of the solubility of a solid with temperature is shown graphically as a solubility curve (Figure 29).

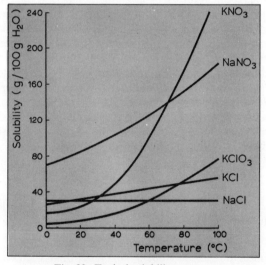

Fig. 29. Typical solubility curves

Discontinuities in the solubility curve are usually seen when (a) a change in the crystal form of the solid takes place, or (b) a change in the degree of hydration of the salt occurs (Figure 30).

The solubility of calcium sulphate decreases with increasing temperature; for example at 40°C the solubility is 0·21, while at 100°C it is 0·17 g/100 gH₂O.

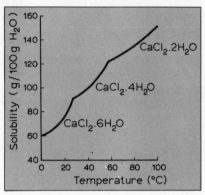

Fig. 30. Discontinuous solubility curve for calcium chloride

Fractional crystallization It is possible to separate the components of a mixture of solids in solution by slowly cooling a hot concentrated solution. The nature of the crystals first deposited depends on the relative concentration of each species and on the solubility of the substances which may be deposited from solution.

For instance, in a solution containing *equimolar* proportions of sodium, potassium, chloride and nitrate ions, sodium chloride would be deposited first as the solution is concentrated by evaporation (see Figure 29). After the removal of these crystals from the hot solution, cooling would give rise to a deposit containing a small quantity of sodium chloride and a large quantity of potassium nitrate. The potassium nitrate could be purified by dissolving the crystals in the minimum quantity of water and recrystallizing.

Example 29 The solubility at 20°C of potassium chlorate is 8 and of potassium chloride 32 g/100 gH₂O. What crystals would deposit first if solutions, initially at 100°C and containing

(a) 0·1 mole potassium chloride, 0·1 mole potassium chlorate and 100 g water
(b) 35 g potassium chloride, 0·06 mole potassium chlorate and 100 g water

were cooled to 20°C?

Solution 1 mole potassium chloride weighs 74·5 g, and 1 mole potassium chlorate weighs 122·5 g. Therefore, solution (a) contains 7·45 g potassium chloride and 12·25 g potassium chlorate per 100 g water. It is clear that the solubility of potassium chlorate is exceeded when the solution is cooled to 20°C, while that of potassium chloride is not. Hence the deposit will consist of potassium chlorate.

Solution (b) contains 35 g potassium chloride and 122·5 × 0·06 or 7·35 g potassium chlorate per 100 g water. In this case the crystalline deposit will consist of potassium chloride, since the solubility of this substance has been exceeded on cooling.

(*Note:* In this calculation, no allowance has been made for the fact that the presence of a salt in solution tends to reduce the solubility of a second salt in the same solution.)

Solutions of Solids in Solids

A pure substance is, with few exceptions, associated with a sharp melting point. In addition, the temperature interval between the point at which a solid first becomes fully molten (melting point) and the point at which a liquid just becomes completely solid (freezing point) is narrow. When an impurity is added to the solid, the melting point alters (it is usually lowered) and the melting range is extended.

Figure 31 shows the temperature–concentration diagram for mixtures of gold and silver. The upper curve (*liquidus*) indicates the

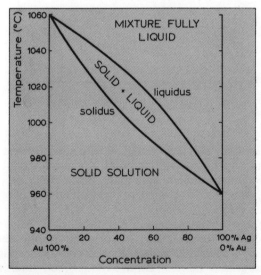

Fig. 31. Temperature–concentration diagram for solutions of gold and silver

temperature at which the mixture becomes completely molten, while the lower curve (*solidus*) shows the temperature to which the molten mixture has to be cooled in order to bring about complete solidification. A molten mixture of gold and silver is a true solution, since the liquid metals are fully miscible. In such circumstances, on cooling, the solid produced does not consist of a mixture of crystals of pure gold and pure silver, but is what is termed an alloy in the case of metals, or more generally, a solid solution. In a solid solution, one component can still be regarded as being dissolved by the other, even in the solid state. Many cases occur where the melting point of

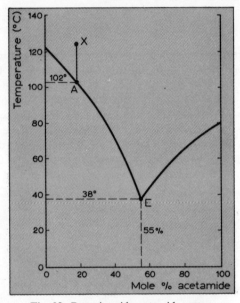

Fig. 32. Benzoic acid–acetamide system

the solid solution falls below that of either component to give a system with a minimum melting point. A typical example is fusion mixture (containing sodium carbonate, m.p. 820°C, and potassium carbonate, m.p. 860°C), which has a minimum melting point of 700°C.

Often, the two components of a system which has a minimum melting point are immiscible in the solid state; this leads to the formation of a eutectic. The benzoic acid–acetamide system shows this behaviour, and the temperature composition graph is given in Figure 32. When a liquid of composition X is cooled, solidification commences at point A (102°C), when crystals of pure benzoic acid separate out. The remaining liquid therefore contains a greater proportion of acetamide, and on further cooling the temperature–composition relation follows the line AE. At E, complete solidification takes place with the deposition of crystals of both pure acetamide and pure benzoic acid. A similar sequence would be noticed when the composition of the liquid mixture is greater than 55% acetamide, but the initial deposit would consist of pure acetamide crystals. E is known as the *eutectic point* for the system, and the temperature at this point, the *eutectic temperature*, is the lowest at which mixtures of these two components can remain liquid.

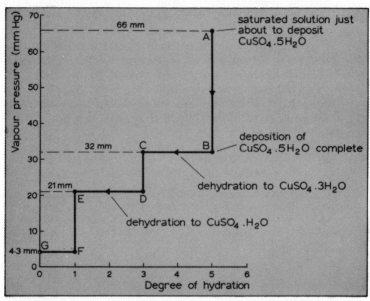

Fig. 33. Vapour pressures of the hydrates of copper (II) sulphate at 40°C

Vapour Pressure of Salt Hydrates

The saturated vapour pressure of water standing over a solution of a salt gradually decreases as the concentration of the solution increases. The vapour pressure of water over a saturated solution of copper (II) sulphate at 40°C is 64 mm of mercury. Removal of this water vapour corresponds to the formation of crystals of the pentahydrate, $CuSO_4.5H_2O$, and the water vapour pressure drops to 32 mm (along the line AB in Figure 33) when crystallization is complete. (This means that the vapour pressure of water at 40°C in equilibrium with solid $CuSO_4.5H_2O$ crystals is 32 mm.) From point B, removal of more water vapour leads to the formation, by dehydration, of an increasing amount of trihydrate, but the vapour pressure standing over the solid stays at 32 mm while any pentahydrate remains. At C, conversion to the trihydrate is complete, and the vapour pressure drops to 21 mm. Further dehydration produces the monohydrate and the anhydrous material, the sequence following the lines DE, EF, FG and GO in Figure 33.

Deliquescence and efflorescence Suppose a salt can exist either in the anhydrous form or as the monohydrate. At 20°C, the water vapour pressure in this country is about 15 mm; consequently, if the change from the monohydrate to the anhydrous salt corresponds to a water vapour pressure above 15 mm, then water will be removed from the monohydrate into the atmosphere and the monohydrate will effloresce. On the other hand, if the monohydrate is deposited from its saturated solution at a vapour pressure below 15 mm, the monohydrate will take up moisture from the air and the hydrate will be deliquescent. The different cases are summarized below, assuming that the temperature is 20°C and the vapour pressure of water in the atmosphere is 15 mm.

Salt	Vapour pressure over a mixture of this hydrate and lower hydrate	Vapour pressure over this hydrate and saturated solution	Condition of salt
$CuSO_4.5H_2O$	5 mm	16 mm	stable
$CaCl_2.6H_2O$	2·6 mm	7·6 mm	deliquesces
$Na_2CO_3.10H_2O$	24 mm	26·2 mm	effloresces

The Properties of Dilute Solutions

When a solution contains a *non-volatile* solute,

(a) the vapour pressure of the solution is lowered,
(b) the boiling point is increased,
(c) the freezing point is lowered,
(d) the solution is capable of exerting osmotic effects.

The extent of this behaviour depends on the *number* of solute particles present in solution, rather than on their identity, and these properties are known as *colligative* properties. The laws relating to the colligative properties of solutions containing a non-volatile solute apply to *dilute solutions only*.

Lowering of Vapour Pressure

On page 64, Raoult's law for a liquid mixture was stated mathematically in the form:

$$p_A = x_A P_A^0$$

where p_A is the vapour pressure of component A in the vapour, x_A is the mole fraction of A in the liquid mixture and P_A^0 is the vapour pressure of pure A at the given temperature. However, when the solution contains a non-volatile solute, p_A is, in fact, the total vapour pressure of the solution.

Now

$$\frac{p_A}{P_A^0} = x_A = 1 - x_S \quad \text{(where } x_S \text{ is the mole fraction of the solute)}$$

or

$$\frac{P_A^0 - p_A}{P_A^0} = x_S$$

The factor $(P_A^0 - p_A)/P_A^0$ is termed the *relative lowering of vapour pressure*.

Thus, Raoult's law may be stated as follows: *the relative lowering of vapour pressure of a solution is equal to the mole fraction of the solute in the dilute solution.*

By measuring the relative lowering of vapour pressure, the weight of one mole of non-volatile solute can be determined. The apparatus used is shown in Figure 34, and the experiment is illustrated by Example 30.

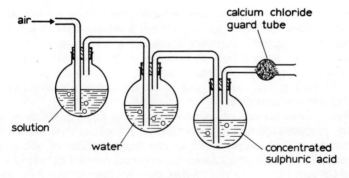

Fig. 34.

Example 30 A solution contains 7·31 g urea in 50 g of water. When air was bubbled through this solution and then successively through water and concentrated sulphuric acid in an apparatus similar to that shown in Figure 34, the bulb containing pure water lost 0·087 g, while a gain in weight of 2·036 g was noted for the sulphuric acid bulb. What is the weight of one mole of urea?

Solution

Loss in weight of solution $\propto p_A$
Loss in weight of water bulb $\propto P_A^0 - p_A \propto 0.087$
Gain in weight of sulphuric acid bulb $\propto P_A^0 \propto 2.36$

Hence,
$$\frac{P_A^0 - p_A}{P_A^0} = \frac{0.087}{2.036} = x_S = 0.043$$

If M is the weight of one mole of urea, then:

$$\text{number of moles urea present in solution} = \frac{7.31}{M}$$

and

$$\text{number of moles water in solution} = \frac{50}{18} = 2.78$$

Therefore,
$$x_S = \frac{7.31/M}{2.78 + 7.31/M} = 0.043$$

From which,
$$M = 58.9$$

Elevation of Boiling Point

On page 55, it was shown that a liquid boils when its saturated vapour pressure is equal to the external pressure. Thus, a solution containing a non-volatile solute which has a lower vapour pressure than the pure solvent boils at a higher temperature than the pure solvent. This can be seen from Figure 35, in which the curve A is for the pure solvent, and B and D are the curves for solutions of increasing concentration.

From these curves, AC is proportional to $P_A^0 - p_A$, which is, in turn, proportional to the concentration of solute in solution (1). Similarly, AE is proportional to the concentration of solute in solution (2). For dilute solutions, the distances AB and BD are small, and CB and ED can be regarded as straight lines, so that ABC and ADE are similar triangles. Thus,

$$\frac{AB}{AD} = \frac{T_1 - T_0}{T_2 - T_0} = \frac{AC}{AE}$$

$$= \frac{\text{lowering of v.p. for solution (1)}}{\text{lowering of v.p. for solution (2)}}$$

or, *the elevation of boiling point is proportional to the concentration of the solution.*

The elevation in boiling point (denoted by ΔT) produced by one mole of solute (which does not associate or dissociate) contained in 100 g of a solvent is termed the *elevation of boiling point constant*, or *molar elevation* for the solvent, and this is denoted by K.

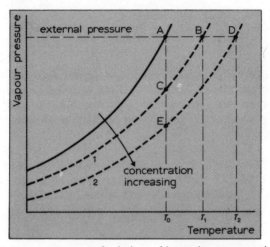

Fig. 35. Vapour pressures of solutions of increasing concentration

Now, if one mole (M grammes) of solute dissolved in 100 g of solvent produces an elevation of K degrees, then, one g of solute dissolved in 100 g solvent would produce an elevation of

$$\frac{1}{M} K \text{ degrees}$$

and w grammes of solute dissolved in 100 g of solvent would produce an elevation of

$$\frac{w}{M} K \text{ degrees}$$

Thus,

$$\Delta T = \frac{w}{M} K$$

It is possible to calculate values of K from the equation

$$K = \frac{0 \cdot 02 \, T^2}{l}$$

where T is the normal boiling point of the solvent, and l is the latent heat of vaporization (in calories) per gramme of solvent. For water, T is 373 K and l is 540·5 cal g^{-1}, so that

$$K = \frac{0 \cdot 02 \times 373^2}{540 \cdot 5} = 5 \cdot 15$$

Values of K for other solvents are:

        benzene 26·7   ether 21·0
        acetone 16·7   chloroform 36·5

Note that, occasionally, the values of K are given for a solution containing one mole of solute in 1000 g of solvent—this is a more dilute solution than the one used in the previous considerations, and so the K values for 1000 g of solvent are ten times *smaller* than those given above.

Example 31 Acetone boils at 56·38°C, and a solution of 1·41 g of an organic solid X in 20 g of acetone boils at 56·88°C. If K for acetone is 16·7, calculate the weight of one mole of X.

Solution If 20 g of acetone contains 1·41 g of X, then 100 g of acetone would contain

$$\frac{1 \cdot 41 \times 100}{20} = 7 \cdot 05 \text{ g of X}$$

Using the relation $\Delta T = (w/M)K$, we obtain

$$M = \frac{7 \cdot 05 \times 16 \cdot 7}{56 \cdot 88 - 56 \cdot 38} = 235$$

Measurement of boiling point elevation To obtain accurate results, the following precautions must be observed:

    (a) an accurate large scale thermometer is required,
    (b) the solute must be as involatile as possible,
    (c) superheating of the solution must be avoided.

A diagram of the apparatus used in *Beckmann's method* is given in Figure 36. The experiment, which takes a considerable time to complete, is carried out in the following stages.

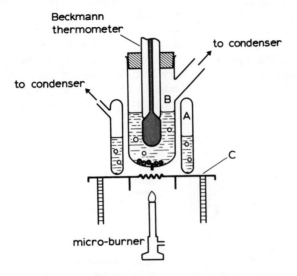

Fig. 36. Beckmann's apparatus for determining boiling point elevations

1. The pure solvent is placed in both the inner tube B, and the annular tube A, each of which is connected to a condenser to return any escaping vapour. Some glass beads are placed in B, and a short length of platinum wire is sealed through the base of the tube in order to eliminate superheating.
2. The Beckmann thermometer is adjusted, and fitted into the central tube. Then the whole unit is heated very gently by means of a micro-burner flame directed at the small gauze insert in the asbestos mat C.
3. Heating is continued until the boiling point of the pure solvent is attained, and the reading on the Beckmann is noted.
4. After cooling, a small weighed sample of solute is added to the inner vessel by means of the side arm. Heating is recommenced, and the boiling point of the solution is found.
5. Further additions of solute are made, and a succession of boiling point elevations are determined. A graph of boiling point elevation against the concentration of the solution is plotted, so that the best average value can be obtained.

The *Beckmann thermometer* is shown in Figure 37. This instrument is not intended to give a reading of temperature but to enable temperature differences to be accurately measured. It is used over a very small range of temperature and the scale covers five or six Centigrade degrees. Hence, before use, it must be set so that the mercury level at the temperature of the experiment comes near to the centre of the scale. To do this, the thermometer bulb is warmed until the mercury thread joins up with that in the upper reservoir. A small scale carried on this reservoir shows the point at which the mercury thread should be broken so that the mercury level falls on the main scale at the temperature of the determination; the thread is broken by giving the upper part of the instrument a gentle tap while holding it upright.

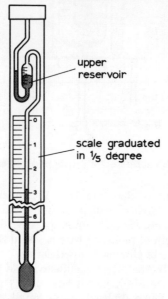

Fig. 37. Beckmann thermometer

The disadvantages of the Beckmann method stem mainly from the problems of eliminating superheating, and the length of time required to complete a determination. In addition, heat is lost from the central tube by radiation, but this is reduced to a minimum by having boiling solvent in the outer annular vessel.

Depression of Freezing Point

At the freezing point, the vapour pressure of a liquid and that of the pure solid become equal. It follows that the lowering of vapour pressure caused by the presence of a non-volatile solute in a solution leads to the solution freezing at a temperature below that of the pure solvent (Figure 38).

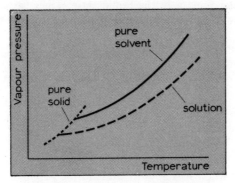

Fig. 38. Depression of freezing point

Again it can be shown that, for dilute solutions, the extent of the freezing point depression is proportional to the concentration of the solution. The depression of freezing point (in Centigrade degrees) brought about by dissolving one mole of non-volatile solute in 100 g of solvent, is termed the *depression of freezing point constant*, or *molar depression*, and it is again denoted by K. Values of K may be determined experimentally, or from the equation

$$K = \frac{0 \cdot 02 \; T^2}{l}$$

where T is the freezing point of the solvent and l the latent heat (in cal) of fusion per gramme of solvent. Some values of K are:

water	18·6	acetic acid	39·3
benzene	50·0	camphor	400

If 1 mole (M g) of solute dissolved in 100 g solvent cause a depression of K degrees, then w g of solute dissolved in 100 g of solvent would cause a depression of

$$\frac{w}{M} \; K \text{ degrees}$$

and, as before,

$$\Delta T = \frac{w}{M} K$$

Example 32 250 g of water contains 0·1 mole of a non-volatile solute. Calculate the quantity of ice deposited when the solution is cooled to $-3\cdot72°C$. K for water (100 g) is 18·6.

Solution When the freezing point of the solution reaches $-3\cdot72°C$, let x g of water remain in the liquid state. Then, at $-3\cdot72°C$, x g of water dissolve 0·1 mole solute, and 100 g of water would dissolve

$$\frac{0\cdot1 \times 100}{x} \text{ mole solute} \left(= \frac{w}{M}\right)$$

$$\Delta T = 3\cdot72 = \frac{0\cdot1 \times 100}{x} \times 18\cdot6$$

Therefore

$$x = \frac{0\cdot1 \times 100 \times 18\cdot6}{3\cdot72} = 50 \text{ g}$$

Since 250 g of water was initially present, the weight of ice deposited is 200 g.

Measurement of the depression of freezing point Accurate results are easier to obtain in these determinations than in measuring the elevation of boiling point, but precautions must be taken to avoid super-cooling.

The apparatus used in *Beckmann's method* consists of a tube fitted with a side arm, stirrer and Beckmann thermometer; it is surrounded by an air jacket and immersed in a well-stirred freezing mixture, as shown in Figure 39.

1. A known weight of pure solvent is placed in the inner tube A, which is cooled to a temperature just above the freezing point of the solvent.
2. The apparatus is assembled as shown, and the solvent is thoroughly stirred to eliminate super-cooling.
3. The steady temperature corresponding to the freezing point of the pure solvent is noted. If super-cooling has taken place, the temperature will fall initially below the freezing point of the pure solvent, but will rise to assume the steady value of the freezing point as the solvent crystallizes. If a considerable degree of super-cooling has taken place, it is better to repeat the determination.
4. A weighed quantity of solute (in pellet form if possible) is added to the solvent via the side arm in tube A. The freezing point of the solution is then determined.
5. Further weighed portions of solute are added and the freezing point noted after each addition. From a graph of the results, the best average value is taken.

Rast's micro method depends on the fact that camphor has the very high value of 400 for the cryoscopic constant, so that a small quantity of a solute soluble in camphor can produce a considerable depression in the freezing point, thus eliminating the need for a Beckmann thermometer. The freezing point of pure camphor is first determined and then the freezing point of a mixture containing known weights of camphor and solute is found. A thermometer graduated in tenths of a degree is suitable for this determination.

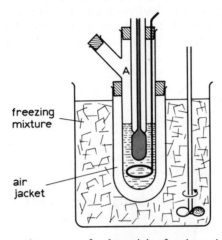

Fig. 39. Beckmann's apparatus for determining freezing point depressions

Example 33 Deduce the weight of one mole of naphthalene if a mixture of 5 g camphor and 0·16 g naphthalene freezes at 167·3°C. (The freezing point of pure camphor is 177·3°C, and K for 100 g camphor is 400.)

Solution Using $\Delta T = (w/M) K$, we find that the weight w of napthalene in 100 g camphor is given by

$$w = \frac{100 \times 0\cdot16}{5} = 3\cdot2 \text{ g}$$

Also,

$$M = \frac{3\cdot2 \times 400}{177\cdot3 - 167\cdot3} = 128$$

Osmosis

Osmosis is the term used to describe the diffusion of a solvent from a dilute to a more concentrated solution through a semi-permeable membrane. (A semi-permeable membrane is a barrier which allows the passage of small solvent molecules, but does not allow larger solute molecules to pass through its pores.) In Figure 40, a concentrated solution is contained in a tube sealed with a semi-permeable membrane. When this tube is immersed in the pure solvent, the solvent molecules diffuse into the concentrated solution, raising the liquid level in the tube and creating a hydrostatic head. Once the hydrostatic head has been created, the solvent molecules passing through the semi-permeable membrane do so against an increasing hydrostatic pressure. As time passes, the liquid level in the tube rises progressively more slowly and eventually ceases to rise altogether.

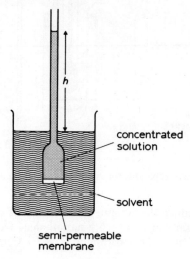

Fig. 40. Osmosis

The hydrostatic head h which is established between a solution and the pure solvent is known as the osmotic pressure (Π) of the solution. The osmotic pressure of a solution is thus an expression of the ability of a solution to produce this hydrostatic head of pressure under these conditions. At the final stage of the experiment, the liquid level does not rise or fall, although solvent molecules are still diffusing through the semi-permeable membrane. This is due to the fact that they are diffusing into, and out of, the concentrated solution at the same rate.

Alternatively, the osmotic pressure of a solution can be regarded as the minimum external pressure which will prevent the formation of a hydrostatic head between a solution and the pure solvent when they are separated by a semi-permeable membrane. Solutions which have the same osmotic pressure are said to be *isotonic*.

Measurement of osmotic pressure: Berkeley and Hartley's method
The apparatus is shown in Figure 41; it consists of an inner porous tube carrying a semi-permeable membrane and containing the pure solvent, surrounded by a steel outer tube which contains the solution, and a device for the application and measurement of external pressure.

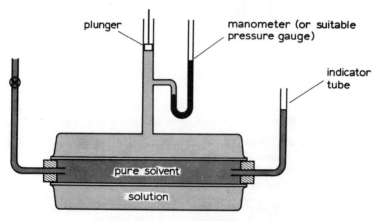

Fig. 41. Berkeley and Hartley's apparatus

One side arm attached to the central tube of this apparatus is used for filling purposes, while the other tube serves to indicate any rise or fall in the liquid level. The applied pressure, which may be of the order of several atmospheres, is adjusted until the liquid level in the indicator tube neither rises nor falls; it is then equal to the osmotic pressure of the solution. The semi-permeable membrane is produced within the pores of the inner tube before the start of the experiment. The inner tube is filled with a solution of copper (II) sulphate (either *in situ*, or withdrawn from the apparatus and standing in a beaker; see Figure 42), and surrounded with a solution of potassium hexacyano-ferrate (II). Electrodes are inserted into the solutions (the anode dips into the copper solution) and an electric current is passed, causing the formation of a deposit of copper hexacyano-ferrate (II) within the porous tube; this deposit acts as a semi-permeable membrane.

The laws of osmotic pressure W. F. Pfeffer in 1877 measured the

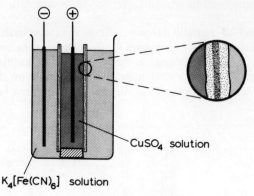

Fig. 42. Production of copper hexacyanoferrate (II) membrane

osmotic pressure Π of a series of sucrose solutions at different concentrations and temperatures. From these results, it was apparent that:

(a) the osmotic pressure of a solution is proportional to the absolute temperature,
(b) the osmotic pressure of a solution is proportional to the concentration of the solution.

J. H. van't Hoff pointed out the analogy between ideal gases and dilute solutions: as

$$\Pi \propto \text{concentration},$$

and as the concentration in moles litre^{-1} is the reciprocal of the volume, V litres, containing one mole, it follows that $\Pi \propto 1/V$ which compares with Boyle's law.

The fact that $\Pi \propto T$ is the equivalent of Charles' law. Combining Boyle's and Charles' laws leads to the equation

$$PV = nRT$$

for n moles of ideal gas, and similarly for a dilute solution,

$$\Pi V = nkT$$

for n moles of solute. By experiment, it was found that a 0·029 M solution of sucrose gave an osmotic pressure of 0·701 atm at 20°C. Substituting these values into $\Pi V = nkT$,

$$k = \frac{\Pi V}{nT} = \frac{0\cdot701 \times 1}{0\cdot029 \times 293}$$

$$= 0\cdot0825 \text{ litre atm deg}^{-1} \text{ mole}^{-1}$$

which is very close to the value of R (0·0821 litre atm deg^{-1} mole^{-1}) used in the ideal gas equation. Consequently, the equation

$$\Pi V = nRT$$

is used in connection with osmotic pressure calculations. The following data shows the osmotic pressure (measured by Pfeffer) of a sucrose solution compared with the calculated ideal gas pressure at the same temperature.

Temperature °C	Osmotic pressure of 0·029 M sucrose	Calculated ideal gas pressure
6·8	0·664	0·666
15·5	0·684	0·687
36·0	0·746	0·737

The agreement between these figures confirms that the ideal gas law holds for dilute solutions, and a conclusion similar to Avogadro's hypothesis can be drawn. That is, equal numbers of different solute molecules when dissolved in the same volume of solvent produce the same osmotic pressure at the same temperature.

Example 34 What is the osmotic pressure of a solution containing 0·1 mole of non-volatile solute in 100 cm^3 of solution at 27°C?

Solution Using $\Pi V = nRT$, we obtain

$$\Pi \times \frac{100}{1000} = 0 \cdot 1 \times 0 \cdot 082 \times 300$$

from which

$$\Pi = 24 \cdot 6 \text{ atm}$$

Association and Dissociation

In common with vapour pressure, boiling point elevation and freezing point depression, the osmotic pressure of a solution depends on the number (and not on the nature) of the solute particles in solution. However, when placed in solution a solute may:

1. *Associate*—i.e. form aggregates which contain more than one molecule. Consequently, the number of particles in solution is decreased, and the extent of the measured colligative property is reduced. For example, benzoic acid is associated into double molecules in benzene. Therefore, 2 moles of benzoic acid must be dissolved in 100 g of benzene to produce a depression equal to K (50 degrees).
2. *Dissociate* either partly or completely. For instance, one mole of

sodium chloride which consists of one mole of sodium ions and one mole of chloride ions gives two moles of 'particles' in solution. Hence, 0·5 mole of sodium chloride dissolved in 100 g of water produces a depression of freezing point equal to K (18·6 degrees).

To express this behaviour mathematically, Van't Hoff introduced the factor i, where

$$i = \frac{\text{observed value}}{\text{calculated value}}$$

If i is greater than unity, dissociation has taken place, while an i value of less than one corresponds to association of the solute.

Example 35 A solution containing 7·5 g of potassium bromide per litre has an osmotic pressure of 3 atmospheres at 18°C. What deductions can be made from this result?

Solution Using $\Pi V = nRT$,

$$3 \times 1 = n \times 0{\cdot}082 \times 291$$

Thus n, the number of moles apparently present, is given by

$$n = \frac{3}{0{\cdot}082 \times 291} = 0{\cdot}126$$

But the number of moles of potassium bromide actually present is

$$\frac{7{\cdot}5}{119} = 0{\cdot}063$$

Thus the Van't Hoff factor is

$$i = \frac{0{\cdot}126}{0{\cdot}063} = 2$$

which means that in solution two moles of ions are present (K^+ and Br^-) for each mole of potassium bromide dissolved.

EXERCISES

1. The saturated vapour pressure (in atmospheres) of benzene at various temperatures between 0 and 100°C is given in the following table:

Temperature	21·8	43·0	61·0	73·2	90	100
Sat. vapour pressure	0·13	0·26	0·53	0·79	1·32	1·76

   Plot these results on a graph, and deduce the boiling point of

benzene at (a) 76 cm, (b) 57 cm pressure. What is the external pressure when the boiling point of benzene is 94°C?
2. 12·33 litres of dry air, measured at 40°C and 744 mm pressure, were passed through a water bulb maintained at 40°C. If the loss of water from the bulb amounted to 0·6514 g, calculate the saturated vapour pressure of water at 40°C.
3. If air can be regarded as a mixture containing 80% by volume of nitrogen, and 20% by volume of oxygen, show that the percentage of oxygen, by volume, in dissolved air is 33·3, given that oxygen is twice as soluble in water as nitrogen.
4. Hydrogen bromide was passed through 10 cm^3 of water at 0°C and 760 mm pressure, until a saturated solution was formed. This solution was diluted to 250 cm^3 with de-ionized water, and 25 cm^3 of the dilute solution gave a precipitate of 0·5037 g of pure dry silver bromide on treatment with excess silver nitrate solution. Calculate the volume of hydrogen bromide dissolved by 1 cm^3 of water at s.t.p.
5. If 27·5 cm^3 of water dissolves 0·6 cm^3 of nitrogen at 7°C, calculate the absorption coefficient of nitrogen for this temperature.
6. The following mixtures of p-bromotoluene and thymol commence to freeze at the temperatures given:

Temperature (°C)	49	39·5	28·5	15·2	14·2	18·8	22·8	26·2
p-bromotoluene (weight %)	0	20	40	60	70	80	90	100

Plot these data, and deduce the freezing temperature and composition of the eutectic mixture.
7. Calculate the total vapour pressure standing over a mixture containing 5 g benzene and 10 g toluene at 52°C. The vapour pressures of pure benzene and pure toluene at 52°C are 30 cm and 10 cm respectively.
8. The following data refer to the glycerol–m-toluidine system, which shows an upper and a lower critical solution temperature. Express the results graphically, and deduce the temperature range over which a mixture containing 75% m-toluidine separates into two layers.

Temperature (°C)	6	10	20	30	50	70	90	110	115	120
m-toluidine (%)	46	28	20	16	14	15	18	25	30	47
		70	82	86	87	86	82	68	62	

9. 40·4 g of a solution of copper (II) sulphate saturated at 20°C was made up to 250 cm^3 with distilled water. 25 cm^3 of the diluted

solution produced 1·012 g of barium sulphate precipitate when treated with a slight excess of barium chloride solution. Calculate the solubility of copper (II) sulphate (as g $CuSO_4$ per 100 g water) at 20°C.

10. The vapour pressure of water at 76°C is 30 cm. A solution containing 180 g of water and 36 g of glucose has a vapour pressure of 29·4 cm. What is the weight of one mole of glucose?

11. Plot a graph relating to the partial miscibility of iso-butyric acid and water, from the following data:

Temperature (°C)	0	5	10	15	20	22	24	24·5
Iso-butyric acid (% weight)	16	16·5	18·0	19·5	22·5	24·5	29·5	38
	80	74·5	69·0	64·0	58·0	54·5	46·5	

From the graph, determine the compositions of the two liquid layers formed when a mixture containing 60% of water and 40% of iso-butyric acid is allowed to settle out at 12·5°C.

12. If 1 g of a substance X, when dissolved in 20 g of water, has a freezing point of $-1·55°C$, calculate the vapour pressure of this solution at a temperature for which the vapour pressure of pure water is 480 mm. (K for 100 g water is 18·6.)

13. Calculate the boiling point of a solution made by dissolving 10 g of cane sugar in 100 g of water. (The weight of one mole of cane sugar is 342, and K for 100 g of water is 5·2.)

14. Naphthalene melts at 80°C and its latent heat of fusion is 35·3 cal g^{-1}. Deduce the freezing point of a mixture containing 2 g anthracene and 60 g naphthalene, if one mole of anthracene weighs 178 g.

15. 6 g of benzoic acid, when dissolved in 100 g of benzene, gives a freezing point depression of 1·23°C. Calculate the apparent weight of one mole of benzoic acid when dissolved in benzene, and comment on the result. (K for 100 g benzene is 49.)

16. What is the osmotic pressure of an aqueous solution containing 6 g urea, $CO(NH_2)_2$, per litre at 47°C? (R is 0·082 l atm deg^{-1}.)

17. At what temperature does a solution containing 1·9 g of magnesium chloride dissolved in 60 g of water freeze? (K for water is 18·6.)

18. Pure acetone boils at 56·4°C. Calculate the weight of acetone that distils over from a solution containing 5 g of a non-electrolyte X and 120 g of acetone, by the time the boiling point of the solution reaches 57·4°C. (The weight of one mole of X is 84 g, and K for 1000 g of acetone is 1·68.)

19. What is the concentration of a glucose ($C_6H_{12}O_6$) solution, expressed in g l^{-1}, which has an osmotic pressure of 1·2 atmospheres at 20°C? (R is 0·082 l atm deg^{-1}.)

20. A solution containing 1/7 mole of naphthalene in 125 g of benzene freezes at 0°C, while a solution containing 6 g of acetic acid in 100 g of benzene freezes at 3·1°C. If benzene freezes at 5·6°C, calculate the cryoscopic constant for benzene, and the apparent weight of one mole of acetic acid when dissolved in benzene. Comment on the last result.
21. What volume of gas dissolves when 500 cm^3 of water is shaken with a gas containing 50% nitrogen, 40% oxygen and 10% carbon dioxide, if the absorption coefficients are 0·015 for nitrogen, 0·03 for oxygen and 1·00 for carbon dioxide? Calculate also the percentage composition of the dissolved gas.
22. When 1·0 g of a substance A, whose vapour has a density of 37 relative to hydrogen, is dissolved in 250 g of a solvent, the freezing point of the solution is 16·3°C. When 4·46 g of a second substance B was dissolved in 200 g of the same solvent, the solution froze at 15·6°C. If the freezing point of the pure solvent is 16·5°C, calculate the weight of one mole of B.
23. If a solution of potassium chloride produces an osmotic pressure of 300 cm at 10°C, calculate the freezing point of the solution, given that the cryoscopic constant for 100 g of water is 18·6.
24. Three solutions were made by dissolving 2 g of the following substances in 200 g of water: fructose ($C_6H_{12}O_6$), anhydrous barium chloride, sodium chloride. The respective freezing points of these solutions were −0·103°C, −0·266°C and −0·625°C. What conclusions can be drawn from these results? (K for 100 g water is 18·6.)
25. An organic compound contains C, H and O only. If 0·1016 g of this compound gave 0·295 g of carbon dioxide and 0·0517 g of water on combustion analysis, and a solution of 1·06 g of the compound in 50 g of benzene gave a depression of freezing point of 1 deg C, calculate the molecular formula of the compound. (K for 100 g of benzene is 50.)
26. If a 4·25% solution of a certain substance, Q, is isotonic with a solution containing 10 g of urea per litre, calculate the weight of one mole of Q. (One mole of urea weighs 60 g.)
27. Calculate the osmotic pressure of a solution containing 50 g of sucrose ($C_{12}H_{22}O_{11}$) per litre, at 15°C.

4. Energy Changes in Chemistry

Heat and Energy

The term energy expresses the capacity of a system to do work, and this energy may take several forms, such as the kinetic energy of moving bodies, the potential energy of a tightly coiled spring, heat energy or electrical energy, etc. The energy of the atoms and molecules in a chemical system is made up of kinetic energy of:

(a) bulk movement of molecules (translational energy),
(b) rotation of the molecules (rotational energy),
(c) vibration of atoms about a mean position in a molecule (vibrational energy)

and potential energy due to the attractive and repulsive forces between the various charged particles. The total of all the forms of energy in a chemical system is termed the *internal energy* of the system, and this is given the symbol U. Changes in internal energy are shown as ΔU, where

$$\Delta U = U_{\text{final state of system}} - U_{\text{initial state}}.$$

All forms of energy may be converted into heat, and Joule in his classical experiment measured a value for the mechanical equivalence of heat. Hence, internal energies and other energy values may be quoted in terms of units of heat energy.

Energy units The introduction of SI (*Système Internationale*) units is intended to present an unambiguous set of standard units accepted internationally in all branches of science and engineering. For example, the various units of heat—calories, Btu, C.H.U. and therms—all disappear, since the SI unit in which energy is expressed is the joule (J). To minimize the difficulties resulting from this change, the units used in this book will be mainly SI, together with some non-SI units to help the student to make the transition.

Quantity	SI unit	Symbol
length	metre	m
mass	kilogramme	kg
time	second	s
electric current	ampere	A
temperature (absolute scale)	kelvin	K

Quantity	SI unit	Symbol
molar heat capacity	joule per mole kelvin	J mole^{-1} K^{-1}
frequency	hertz	Hz
energy	joule	J
force	newton	N
pressure	newton per metre^2	N m^{-2}
power	watt	W
electrical charge	coulomb	C
electromotive force	volt	V
resistance	ohm	Ω

Also in this system, a preferred set of fractions and multiples have been drawn up for use in conjunction with SI units:

Fraction	Prefix	Symbol
10^{-1}	deci	d
10^{-2}	centi	c
10^{-3}	milli	m
10^{-6}	micro	μ
10^{-9}	nano	n
10^{-12}	pico	p

Multiple	Prefix	Symbol
10	deka	da
10^2	hecto	h
10^3	kilo	k
10^6	mega	M
10^9	giga	G
10^{12}	tera	T

Of these, the recommended multiples and fractions are those in which the power of ten is divisible by three. For example, the length we call 1 cm is better expressed, in SI terms, as 0·01 m or 10 mm.

The energy changes with which we are most concerned in this chapter are heat, heat capacity and work.

Heat energy in SI units is expressed in joules (or kilo-joules) instead of calories. 1 calorie equals 4·19 J.

Molar heat capacity* in SI units is expressed in joules per mole kelvin. For example, the energy required to raise the temperature of 1 mole of water (18g) through 1 kelvin is 18 calories or $18 \times 4·19 = 75·42$ joules. Thus the molar heat capacity of water is 75·42 J mole^{-1} K^{-1}.

Work is defined as the product of force and the distance moved

---

* The term *specific heat capacity* (not used in this book) is the heat capacity divided by the mass, appropriate SI units being J kg^{-1} K^{-1}.

in the direction of the force. The difference resulting from the introduction of SI units can be seen from the following example.

Example 36 Calculate the work done in lifting a mass of 1 kg vertically through 1 metre, if the value of g is 981 cm s^{-2}.

Solution (a) Using conventional units:

$$\text{Force} = mg = 1000 \times 981 \text{ g cm s}^{-2} \text{ or dynes}$$
$$\begin{aligned}\text{Work done} &= \text{force} \times \text{distance moved}\\ &= 1000 \times 981 \times 100 \text{ dyne cm or erg}\\ &= 981 \times 10^5 \text{ erg}\end{aligned}$$

But 10^7 erg = 1 J, so

$$\text{work done} = 981 \times 10^{-2} = 9.81 \text{ J}$$

(b) Using SI units:

$$\text{Force} = mg = 1 \times 9.81 \text{ newton} = 9.81 \text{ N}$$

(Note that g has the value of 9.81 m s^{-2} in SI units.)

$$\begin{aligned}\text{Work done} = \text{force} \times \text{distance moved} &= 9.81 \times 1\\ &= 9.81 \text{ J}\end{aligned}$$

Internal Energy of a Monatomic Gas

The kinetic theory of gases refers to the random motion of spherical molecules, between which there are no attractive forces. The noble gases approach most closely to this ideal picture. For such a gas, the kinetic theory develops the equation:

$$PV = \tfrac{1}{3}Nmc^2$$

for one mole of gas, where N is the number of molecules, m the mass of each molecule and c their average velocity.

But for one mole of ideal gas,

$$PV = RT$$

so that

$$RT = \tfrac{1}{3}Nmc^2$$

or

$$Nmc^2 = 3RT \qquad (a)$$

If we consider the total internal energy U of this ideal monatomic gas as being made up purely of kinetic energy, then

$$U = \tfrac{1}{2} Nmc^2$$

or

$$Nmc^2 = 2U \qquad (b)$$

Comparing equation (a) with (b), we see that

$$U = \tfrac{3}{2} RT$$

On differentiating,

$$\frac{dU}{dT} = \tfrac{3}{2} R$$

Now dU/dT is the rate of change of the internal energy with temperature, or the amount of heat energy absorbed corresponding to a rise of 1 K in temperature, which is the molar heat capacity of the gas. Hence, for an ideal monatomic gas, the molar heat capacity at constant volume (denoted by C_v) is

$$\tfrac{3}{2} R = \tfrac{3}{2} \times 8\cdot 31 = 12\cdot 46 \text{ J mole}^{-1}\text{K}^{-1}$$

(The measured molar heat capacity of argon at constant volume is $12\cdot 47$ J mole^{-1}K^{-1}, for example.)

Heating Gases: The First Law of Thermodynamics

Suppose it is possible to trap one mole of an ideal gas in a cylinder, by means of a leak-proof, weightless piston, which can move in the cylinder without loss of energy due to friction (Figure 43). The

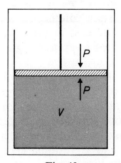

Fig. 43

pressure above and below the piston is P, and the volume occupied by the gas is V. Heat is now given to the gas in the cylinder, which results in an increase in the kinetic energy of the gas molecules. This shows up as a rise in temperature. The molecules, now having a greater kinetic energy, strike the piston with a greater force, so that either:

(a) the piston is pushed upwards—i.e. the gas expands corresponding to a rise in temperature although the pressure remains constant, or,

(b) the piston is clamped to prevent it moving, so that the pressure of the gas in the cylinder increases and the volume remains constant.

The first experiment is said to be conducted at constant pressure, the second at constant volume. In the first experiment, the piston has been raised by the expanding gas, that is, external work has been done; and in both experiments, the internal energy of the gas has increased. Therefore, the heat given to the gas has been changed into an increase in internal energy, and (in one case) into external work. In symbols,

$$q = \Delta U + w$$

where q = heat energy given to system, ΔU = increase in internal energy of system, w = external work done by system. This is a consequence of the *first law of thermodynamics* (often called the law of conservation of energy) which states that *energy cannot be created or destroyed, but may change its form*.

The Molar Heat Capacities of Gases

The molar heat capacity of a gas is defined as the heat required to raise the temperature of one mole of gas by one degree. From the preceding section, it follows that a gas has two values of molar heat capacity, one at constant volume, C_v, and the other at constant pressure, C_p; in the second case external work is performed at the expense of some of the heat given to the gas.

Heat capacity at constant volume The molar heat capacity for an ideal monatomic gas has been shown on page 105 to be $\frac{3}{2}R$ (at constant volume). Molar heat capacities for polyatomic gases (i.e. gases which contain more than one atom per molecule) are greater than $\frac{3}{2}R$, since with these gases the heat energy absorbed is used to increase the rotational and vibrational energy of the molecules in addition to increasing the kinetic energy of motion of the molecules. Consequently, the value of C_v for these gases rises with increasing complexity of the gas molecule. For example, the molar heat capacity at constant volume for argon is 12·47 J mole^{-1}K^{-1}, for carbon monoxide it is 20·5 J mole^{-1} K^{-1}, and for carbon dioxide it is 27·7 J mole^{-1} K^{-1}.

Heat capacity at constant pressure Under these conditions, the heat energy absorbed causes an increase in the internal energy of the gas and an expansion. Thus,

$$C_p = C_v + \text{work done}$$

In order to calculate a value for C_p, the amount of energy expended in performing external work must be deduced.

Work of expansion of a gas at constant pressure Suppose one mole of ideal gas is contained in a cylinder of area of cross-section A, which is fitted with a weightless, frictionless piston (Figure 44). Initially, the piston is in position (1). The pressure above the piston and in the cylinder is P. As a result of some change in the system—a chemical reaction or the absorption of a small quantity of heat—the gas expands slightly so that the piston is pushed upwards by a distance d to position (2).

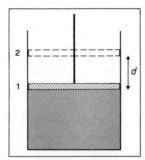

Fig. 44. Work of expansion of a gas

$$\begin{aligned}
\text{External work } w &= \text{force} \times \text{distance moved} \\
&= (P \times A) \times d \quad \text{(since pressure is force per unit area)} \\
&= P \times (A \times d) \\
&= P \times \Delta V
\end{aligned}$$

(where ΔV is the increase in volume occupied by the gas).

Example 37 Calculate the work done when one mole of ideal gas, at a pressure P and at T K expands due to a rise of 1 K in temperature, the pressure remaining constant.

Solution Since 1 mole ideal gas is present, $PV = RT$, so

$$V = \frac{RT}{P} \quad \text{at } T \text{ K}$$

At $(T+1)$ K,

$$\text{Volume of gas} = \frac{R(T+1)}{P}$$

Thus
$$\Delta V = \text{final volume} - \text{initial volume}$$
$$= \frac{R}{P}(T+1-T) = \frac{R}{P}$$

Hence,
$$\text{work done} = P \times \Delta V = P \times \frac{R}{P} = R$$

The general result obtained in Example 37 can be used in connection with the deduction of C_p, since this is the work done when one mole of gas expands following a rise of 1 K in temperature.
Thus,
$$C_p = C_v + \text{work done}$$
$$= C_v + R$$

For an ideal, monatomic gas, $C_v = \tfrac{3}{2}R$ and $C_p = C_v + R = \tfrac{5}{2}R$. For diatomic gases, it can be shown that the theoretical value of C_v is $\tfrac{5}{2}R$, which gives a value of $\tfrac{7}{2}R$ for C_p.

The ratio of the molar heat capacities of a gas, C_p/C_v or γ, is thus:

for a monatomic gas,
$$\gamma = \frac{\tfrac{5}{2}R}{\tfrac{3}{2}R} = \tfrac{5}{3} = 1\cdot 67$$

for a diatomic gas,
$$\gamma = \tfrac{7}{5} = 1\cdot 4$$

and for triatomic gases, $\gamma = 1\cdot 33$. For polyatomic gases, γ is less than 1·33 but greater than unity.

Values of γ may be obtained from measurements other than the direct determination of the two molar heat capacities of a gas, and in this way some knowledge of the atomicity of the gas molecule is obtained. Some experimental values are:

Gas	C_p (J mole^{-1} K^{-1})	C_v (J mole^{-1} K^{-1})	γ
helium	20·9	12·6	1·66
hydrogen	28·7	20·4	1·41
oxygen	29·2	20·9	1·40
carbon dioxide	36·8	27·7	1·29
ammonia	36·6	28·0	1·31
ethene	42·8	34·2	1·25

The heat energy given to a system at constant volume is used exclusively to increase the internal energy of the system; i.e.,

$$\text{heat absorbed at constant volume} = \Delta U$$

But, at constant pressure, the heat energy given to the system not only increases the internal energy but leads to the performance of external work, due to a change in volume. Thus,

$$\text{heat absorbed at constant pressure} = \Delta U + P \Delta V$$
$$= \Delta H$$

The heat absorbed at constant pressure is said to cause an increase in the *heat content H*, or *enthalpy*, of the system. Since most chemical systems are studied under conditions of constant pressure, the heat changes accompanying any change in the system are changes in heat content, and are given as ΔH values.

Heats of Reaction

The heat content change corresponding to a particular reaction is written alongside the equation for the reaction; for example,

$$Fe + 2HCl = FeCl_2 + H_2 \quad \Delta H = -84 \text{ kJ}$$

Such equations are known as *thermochemical* equations. Since changes in the property of a system are reckoned as (final value − initial value), a negative value for ΔH shows that the system has lost energy. In consequence, *heat is evolved* as the reaction progresses, and the reaction is said to be *exothermic*. *Endothermic* reactions are those in which heat is taken in, that is ΔH has a positive value.

The ΔH value in a thermochemical equation is appropriate to the number of moles of each substance taking part in the reaction as shown by the equation. Therefore, should two moles of iron react with four moles of hydrochloric acid, 168 kJ of heat would be evolved.

Example 38 Calculate the amount of heat liberated when 20 g of sodium hydroxide reacts with an excess of sulphuric acid:

$$2NaOH + H_2SO_4 = Na_2SO_4 + 2H_2O \quad \Delta H = -113 \cdot 4 \text{ kJ}$$

Solution When 2×40 g of sodium hydroxide react, 113·4 kJ of heat is evolved. Therefore, when 20 g of sodium hydroxide reacts with excess sulphuric acid, the heat evolved is

$$\frac{113 \cdot 4 \times 20}{2 \times 40} = 28 \cdot 35 \text{ kJ}$$

The heat evolved or absorbed during a reaction also depends on the physical state of the products and reactants, and on the external conditions. For example,

$$H_2 + \tfrac{1}{2}O_2 = H_2O(\text{liq}) \quad \Delta H = -285 \text{ kJ at } 25°C$$

But,
$$H_2 + \tfrac{1}{2}O_2 = H_2O(\text{gas}) \quad \Delta H = -246 \text{ kJ at } 125°C$$

Consequently, in the following definitions corresponding to particular types of reactions, the reactants and products are assumed to be in their normal states at 298 K and one atmosphere pressure, and heat changes appropriate to these standard conditions are designated ΔH^0. The subscripts s, l and g refer to the solid, the liquid and the gaseous states respectively.

Heat of combustion This is the heat evolved when one mole of a substance is completely burnt in oxygen, corrected to the standard conditions outlined above. For example,

$$C_{s,\text{ graphite}} + O_{2,g} = CO_{2,g} \quad \Delta H_c^0 = -405 \text{ kJ}$$

Heat of formation This is the heat change accompanying the formation of one mole of a compound from its elements under standard conditions. For example,

$$C_{s,\text{ graphite}} + 2H_{2,g} = CH_{4,g} \quad \Delta H_f^0 = -74 \cdot 3 \text{ kJ}$$

Note that this definition applies to the formation of *one mole of product*, and in thermochemical equations, fractional numbers of molecules may be used to accord with this definition. For example,

$$\tfrac{1}{2}H_{2,g} + \tfrac{1}{2}Cl_{2,g} = HCl_g \quad \Delta H_f^0 = -92 \cdot 4 \text{ kJ}$$

Sometimes, heats of combustion and heats of formation relate to the same reaction; for instance, the heat of formation of carbon dioxide is the same as the heat of combustion of graphite.

Heat of neutralization This is the heat evolved when one mole of hydrogen ion (from an acid) is just neutralized by a base in dilute solution. For example,

$$HCl + NaOH = NaCl + H_2O \quad \Delta H = -57 \cdot 3 \text{ kJ}$$

Hess's law (the law of constant heat summation) *The heat change accompanying a chemical reaction depends only on the final and initial states of the system, and is independent of all intermediate states.*

This means that it is possible to determine the heat change for a reaction which does not take place directly, by summing up a succession of intermediate reactions. This is illustrated in Figure 45.

For example, the heat of formation of carbon dioxide is given by

$$C + O_2 = CO_2 \quad \Delta H = -404 \cdot 8 \text{ kJ}$$

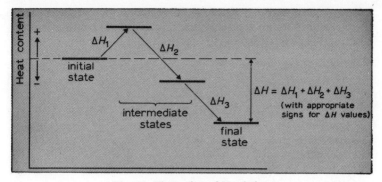

Fig. 45. Illustration of Hess's law

This reaction could be carried out in the following two stages:

$$C + \tfrac{1}{2}O_2 = CO \quad \Delta H_1 = -109.2 \text{ kJ} \quad (1)$$
$$CO + \tfrac{1}{2}O_2 = CO_2 \quad \Delta H_2 = -295.6 \text{ kJ} \quad (2)$$

Adding equations (1) and (2),

$$C + O_2 + CO = CO + CO_2$$

or, since the CO cancels on each side of the equation,

$$C + O_2 = CO_2 \quad \Delta H = \Delta H_1 + \Delta H_2 = -404.8 \text{ kJ}$$

Thermochemical equations Since these equations may be added, subtracted, multiplied and divided like ordinary alegbraic equations, the calculation of the heat of reaction is possible for a case where direct measurements cannot be made.

Example 39 The heats of formation of carbon dioxide and water are 395 kJ and 285 kJ respectively, and the heat of combustion of acetic acid is 869 kJ (all exothermic). Calculate the heat of formation of acetic acid.

Solution The data given are expressed in the form of three thermochemical equations.

$$C + O_2 = CO_2 \quad\quad\quad\quad\quad\quad \Delta H_1 = -395 \text{ kJ} \quad (1)$$
$$H_2 + \tfrac{1}{2}O_2 = H_2O \quad\quad\quad\quad\quad \Delta H_2 = -285 \text{ kJ} \quad (2)$$
$$CH_3COOH + 2O_2 = 2CO_2 + 2H_2O \quad \Delta H_3 = -869 \text{ kJ} \quad (3)$$

The equation corresponding to the formation of acetic acid from its elements is

$$2C + 2H_2 + O_2 = CH_3COOH \quad \Delta H_f = ?$$

This equation may be obtained from the three thermochemical equations on multiplying equations (1) and (2) by two and adding, finally subtracting equation (3):

$2 \times (1)$	$2C + 2O_2 = 2CO_2$
$2 \times (2)$	$2H_2 + O_2 = 2H_2O$
Adding	$\overline{2C + 3O_2 + 2H_2 = 2CO_2 + 2H_2O}$
Subtract (3)	$CH_3COOH + 2O_2 = 2CO_2\ 2H_2O$
Giving, after re-arranging	$2C + O_2 + 2H_2 = CH_3COOH$
$2 \times (1)$	$2 \times \Delta H_1 = -790$ kJ
$2 \times (2)$	$2 \times \Delta H_2 = -570$ kJ
Adding	$\overline{\Delta H = -1360\text{ kJ}}$
Subtract (3)	$\Delta H_3 = -869$ kJ
Giving	$\overline{\Delta H_f = -1360 - (-869) = -491\text{ kJ}}$

Example 40 Using the heats of formation for water and carbon dioxide given in Example 39, and given that the heats of combustion of ethane and ethyne (acetylene) are 1554 kJ and 1302 kJ (both exothermic) respectively, calculate the heats of formation of ethane and ethyne, and deduce the heat change accompanying the reduction of one mole of ethyne to ethane.

Solution The thermochemical equations are:

$$C + O_2 = CO_2 \qquad \Delta H_1 = -395 \text{ kJ} \qquad (1)$$
$$H_2 + \tfrac{1}{2}O_2 = H_2O \qquad \Delta H_2 = -285 \text{ kJ} \qquad (2)$$
$$C_2H_6 + 3\tfrac{1}{2}O_2 = 2CO_2 + 3H_2O \qquad \Delta H_3 = -1554 \text{ kJ} \qquad (3)$$
$$C_2H_2 + 2\tfrac{1}{2}O_2 = 2CO_2 + H_2O \qquad \Delta H_4 = -1302 \text{ kJ} \qquad (4)$$

The equation for the formation of ethane is given by

$$2 \times (1) + 3 \times (2) - (3):$$

$$2C + 2O_2 + 3H_2 + 1\tfrac{1}{2}O_2 - C_2H_6 - 3\tfrac{1}{2}O_2$$
$$= 2CO_2 + 3H_2O - 2CO_2 - 3H_2O$$

which re-arranges to

$$2C + 3H_2 = C_2H_6$$

Similarly,

$$\Delta H = 2 \times \Delta H_1 + 3 \times \Delta H_2 - \Delta H_3$$
$$= -790 - 855 - (-1554)$$
$$= -91 \text{ kJ}$$

The equation for the formation of ethyne is found by

$$2 \times (1) + (2) - (4),$$

giving
$$2C + 2O_2 + H_2 + \tfrac{1}{2}O_2 - C_2H_2 - 2\tfrac{1}{2}O_2$$
$$= 2CO_2 + H_2O - 2CO_2 - H_2O$$

which re-arranges to
$$2C + H_2 = C_2H_2$$

Similarly,
$$\Delta H = 2 \times \Delta H_1 + \Delta H_2 - \Delta H_4$$
$$= -790 - 285 - (-1075)$$
$$= 227 \text{ kJ}$$

These heats of formation are shown in the thermochemical equations (5) and (6):

$$2C + 3H_2 = C_2H_6 \quad \Delta H_5 = -91 \text{ kJ} \tag{5}$$
$$2C + H_2 = C_2H_2 \quad \Delta H_6 = 227 \text{ kJ} \tag{6}$$

The heat of hydrogenation of ethyne to ethane is found by subtracting equation (6) from equation (5):

$$2C + 3H_2 - 2C - H_2 = C_2H_6 - C_2H_2$$

which re-arranges to give
$$C_2H_2 + 2H_2 = C_2H_6$$
and
$$\Delta H = \Delta H_5 - \Delta H_6 = -91 - 227 = -318 \text{ kJ}$$

In the above example, the formation of ethyne is an endothermic reaction; consequently ethyne is known as an endothermic compound, which is a general term given to those substances which have a positive heat of formation.

The Determination of the Heat of Combustion

The measurement is carried out in a bomb calorimeter, shown in Figure 46, in which the substance under test is burnt in an atmosphere of compressed oxygen.

Briefly, the experiment is performed in the following stages:

1. A weighed amount of the sample (about 0·5 g) is placed in the crucible, and a loop of cotton is positioned so as to act as a fuse between the ignition wire and the sample.
2. The bomb is assembled and charged through the pressure valve to 20 or 30 atmospheres with oxygen.
3. The bomb is then placed in a water calorimeter of known water equivalent, and the whole is surrounded by a jacketed vessel which reduces heat losses by radiation (Figure 46b).

4. The temperature of the water in the calorimeter is taken (using an accurate thermometer graduated in tenths of a degree) at regular intervals to allow for the application of a cooling correction.
5. The bomb is fired, and the temperature of the water in the calorimeter is taken at intervals, continuing for some minutes after the maximum temperature has been attained.
6. A full cooling correction is applied. If the heat capacity of the bomb and water calorimeter is known, the heat evolved during the combustion of the sample can be calculated. (A correction is applied for the heat liberated by the combustion of the cotton loop.) The heats of combustion found in this way are ΔE values, and these are converted into ΔH values by allowing for the external work of expansion.

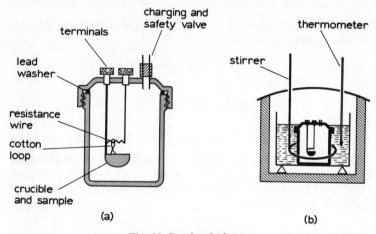

Fig. 46. Bomb calorimeter

Heat of atomization This is defined as the energy required to convert one mole of an element from its normal state at 298 K and one atmosphere pressure into free atoms.

Heats of atomization are difficult to measure (they are usually obtained from spectroscopic data), but they are useful in relation to bond energies. For example:

$$H_{2,g} = 2H_{atoms} \qquad \Delta H = 433 \text{ kJ} \qquad (1)$$
$$O_{2,g} = 2O_{atoms} \qquad \Delta H = 495 \text{ kJ} \qquad (2)$$
$$H_{2,g} + \tfrac{1}{2}O_{2,g} = H_2O_g \qquad \Delta H = -246 \text{ kJ} \qquad (3)$$

Then $(3)-(1)-\frac{1}{2}\times(2)$ gives:

$$2H_{atoms} + O_{atom} = H_2O \quad \Delta H = -926 \text{ kJ}$$

This energy change corresponds to the formation of two oxygen–hydrogen bonds, so that the average bond energy of an oxygen–hydrogen bond in water is 463 kJ.

Heat of hydrogenation The heat of hydrogenation is the heat change accompanying the conversion of one mole of an unsaturated organic compound into the corresponding saturated compound by hydrogen. Values for heats of hydrogenation can indicate peculiarities in the bonding in unsaturated systems. For instance,

$$C_2H_4 + H_2 = C_2H_6 \quad \Delta H = -126 \text{ kJ}$$

This change relates to

$$\diagup\!\!\!\!C\!=\!C\diagdown + H_2 \rightarrow H-\underset{|}{\overset{|}{C}}-\underset{|}{\overset{|}{C}}-H$$

If benzene contains three double bonds, the heat evolved during the hydrogenation of benzene to cyclohexane would be expected to be

$$\Delta H = 3 \times (-126) = -378 \text{ kJ}$$

The actual value is -210 kJ, the difference being accounted for by the fact that the electrons are delocalized over the benzene molecule, rather than concentrated in the form of three double bonds.

Energy and Bonding

Many of the properties of chemical substances can be related to their structure, and especially to the nature of the forces holding atoms together. A bond is the general term used to describe the union of two atoms, while the force required to separate the united atoms is referred to as bond strength or bond energy. At least four ideal bond types—ionic, covalent, metallic and van der Waals are recognized (see Part I, p. 20) differing in the manner in which the valence electrons serve to unite the atoms.

Ionic bonding In this type of bonding, electrons are transferred from one atom to another (see Part I, p. 20) resulting in the formation of cations and anions. Electrostatic attraction binds the ions in a regular manner to form a crystal lattice; it is therefore wrong to speak of molecules (implying the presence of separate unconnected entities) in relation to ionic compounds.

The process of ion formation may be separated into the following stages for which energy data are available. Taking potassium fluoride as an example:

1. *Production of the cation*
   First stage—formation of individual gaseous metal atoms from the solid.

   $K_s = K_g$ Energy change (per mole) is the heat of atomization of potassium
   crystalline   gaseous
   solid metal   atoms
   $$\Delta H_1 = +90 \text{ kJ}$$

   Second stage—conversion of the gaseous atoms to ions.

   $K_g = K_g^+ + e^-$ Energy change is the first ionization energy of potassium
   $$\Delta H_2 = +415 \text{ kJ}$$

2. *Production of the anion*
   First stage—formation of individual gaseous atoms

   $\frac{1}{2}F_{2,g} = F_g$ Energy change is half the bond dissociation energy of fluorine
   gaseous   gaseous
   molecules   atoms
   $$\Delta H_3 = +77 \text{ kJ}$$

   Second stage—conversion of non-metal atoms to ions

   $F_g + e^- = F_g^-$ Energy change is the electron affinity of fluorine
   $$\Delta H_4 = -349 \text{ kJ}$$

3. *Formation of a crystal lattice from the gaseous ions*

   $K_g^+ + F_g^- = K^+F_s^-$ Energy change is the lattice energy of potassium fluoride
   $$\Delta H_5 = -813 \text{ kJ mole}^{-1}$$

The sum of all these energy changes

$$\Delta H_1 + \Delta H_2 + \Delta H_3 + \Delta H_4 + \Delta H_5 = -580 \text{ kJ mole}^{-1}$$

corresponds to the heat change accompanying the sum of the chemical changes for these stages,

$$K_s + \tfrac{1}{2}F_{2,g} = K^+F_s^- \quad \Delta H_f = -580 \text{ kJ mole}^{-1}$$

That is, it is the heat of formation of potassium fluoride. This sequence is summarized diagrammatically in Figure 47. This series of reactions, which is, in fact, the application of Hess's law to the production of an ionic compound, is known as the *Born–Haber cycle*. Two quantities, electron affinity and lattice energy, used in the cycle require definition.

Electron affinity is defined as the energy (per mole) released when a gaseous atom gains one or more electrons to become a negative ion. This definition implies that a release of energy (which accompanies this process in most cases) should be linked with a positive sign. This is contrary to the general rule adopted for expressing thermochemical quantities (heat evolved by the system is negative) that it is probably better to redefine electron affinity as the energy change accompanying the formation of a negative ion from a gaseous atom. Thus

$$F_g + e^- = F_g^- \quad \Delta H = -349 \text{ kJ mole}^{-1}$$

is the electron affinity of fluorine. Some elements require energy to form an anion:

$$O_g + 2e^- = O_g^{2-} \quad \Delta H = 658 \text{ kJ mole}^{-1}$$

Therefore, when tables of data are consulted, care must be taken to appreciate the sign convention used. (The data given at the end of the book conform with the rule that heat evolved by the system is negative.)

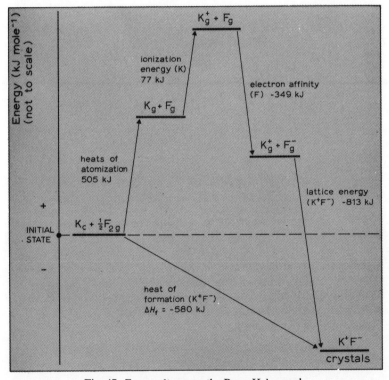

Fig. 47. Energy diagram: the Born–Haber cycle

Lattice energy is defined as the energy, per mole of ionic compound, required to break apart the ions from their stable positions in a crystal lattice and to produce them in an isolated, gaseous state.

It follows, then, that lattice energy gives some measure of the binding force operating in an ionic compound. The calculation of lattice energy from thermochemical data using the Born–Haber cycle is described in Example 41.

Example 41 Calculate the lattice energy of calcium oxide given that the heat of atomization of calcium is 177 kJ mole^{-1}, the bond dissociation energy of oxygen is 500 kJ mole^{-1}, the first and second ionization energies of calcium are 587 kJ mole^{-1} and 1156 kJ mole^{-1} respectively and the heat of formation of calcium oxide is $\Delta H_f = -637$ kJ mole^{-1}. The energy *required* to form one mole O^{2-} ion from gaseous oxygen atoms is 704 kJ.

Solution The stages in the Born–Haber cycle may be constructed as follows:

$$Ca_s = Ca_g \qquad \Delta H_1 = 177 \text{ kJ}$$
$$Ca_g = Ca_g^{2+} + 2e^- \qquad \Delta H_2 = 587 + 1156 \text{ kJ}$$
$$\tfrac{1}{2}O_{2,g} = O_g \qquad \Delta H_3 = \tfrac{1}{2}.500 \text{ kJ}$$
$$O_g + 2e^- = O_g^{2-} \qquad \Delta H_4 = 704 \text{ kJ}$$
$$Ca_g^{2+} + O_g^{2-} = Ca^{2+}O_s^{2-} \qquad \Delta H_5 = ?$$

Also,
$$Ca_s + \tfrac{1}{2}O_{2,g} = Ca^{2+}O_s^{2-} \qquad \Delta H_6 = \Delta H_f = -637 \text{ kJ}$$

Equating,
$$\Delta H_1 + \Delta H_2 + \Delta H_3 + \Delta H_4 + \Delta H_5 = \Delta H_6$$
$$2874 + \Delta H_5 = -637$$

So that ΔH_5, the lattice energy of calcium oxide, is equal to -3511 kJ mole^{-1}.

Significance of lattice energy A knowledge of lattice energies is extremely useful in considering cases involving the breakdown or formation of crystal lattices. (A list of lattice energies and other data is included at the end of the book.)
Typical instances concern:

Melting points of ionic compounds Salts with high lattice energies require a greater input of thermal energy to break apart the crystal lattice. Consequently, such salts have high melting points. The lattice energies and melting points of the sodium halides are compared below:

	Na^+F^-	Na^+Cl^-	Na^+Br^-	Na^+I^-
Lattice energy kJ mole^{-1}	918	779	742	691
Melting point °C	993	801	766	665

Solubility in polar solvents When cations or anions are placed in a polar solvent, an attractive force exists between the ions and the solvent molecules. This results in the clustering of solvent molecules round the ions, a process which releases energy. For a solvent in general, the release of energy is designated as *solvation energy*, while for water in particular, the term *hydration energy* is used. (Hydration energies are not easily determined, but a list of values for some common ions is included at the rear of the book.) In many cases, the solubility of a range of salts runs parallel to the excess of hydration energy of the ions over the lattice energy of the salt. For example, the lattice energies and hydration energies for the silver halides are (values in kJ mole^{-1}):

Salt	Ag^+F^-	Ag^+Cl^-	Ag^+Br^-	Ag^+I^-
Lattice energy (L.E.)	955	901	888	884
Hydration energy (H.E.)	−997	−867	−846	−821
ΔH = L.E.+H.E.	−42	+34	+42	+63

Increasingly insoluble ⟶

Silver fluoride is soluble, while the remainder are insoluble, the solubility product (p. 220) decreasing progressively from chloride to iodide. Not all cases are as clear cut as this, since some salts which absorb heat (i.e. hydration energy less than lattice energy) from the surroundings on dissolving are freely soluble. In these cases, the entropy change (p. 135) accompanying dissolution is significant, and the problem must be discussed in terms of free energy changes (p. 135).

Halogen replacement reactions Fluorine may be substituted for chlorine in a hydrocarbon derivative using an alkali metal fluoride, in a reaction such as:

$$RCH_2Cl + MF = RCH_2F + MCl$$

This exchange reaction involves:

(a) breaking a C—Cl bond and replacing it with a C—F bond. This process liberates about 105 kJ mole^{-1};
(b) converting the chlorine atom produced on the rupture of the above bond into a chloride ion, and at the same time converting a fluoride ion into a fluorine atom—this is a constant factor in all such exchange reactions;
(c) breaking apart a metal fluoride lattice and replacing it by a metal chloride lattice. Reference to the data given at the end of the book shows that the energy required for this process for the alkali metals is:

$$\begin{aligned}
\text{LiF} &\to \text{LiCl} & 185 \text{ kJ mole}^{-1} \\
\text{NaF} &\to \text{NaCl} & 139 \text{ kJ mole}^{-1} \\
\text{KF} &\to \text{KCl} & 105 \text{ kJ mole}^{-1} \\
\text{RbF} &\to \text{RbCl} & 100 \text{ kJ mole}^{-1} \\
\text{CsF} &\to \text{CsCl} & 96 \text{ kJ mole}^{-1}
\end{aligned}$$

Consequently, caesium and rubidium fluorides are the best and lithium fluoride the poorest alkali metal fluoride for use in this type of reaction. If it is intended to replace fluorine by chlorine, the converse will be true. Such conclusions are valuable in problems concerning the synthesis of compounds.

Heats of formation of hypothetical compounds Such heats of formation may be estimated provided we can assume a reasonable value for the lattice energy of the hypothetical compound. The heat of formation of Ca^+Cl^- is calculated from the following Born–Haber cycle.

$$\begin{aligned}
Ca_s &= Ca_g & \Delta H_1 &= 177 \text{ kJ mole}^{-1} \\
Ca_g &= Ca^+_g + e^- & \Delta H_2 &= 587 \text{ kJ mole}^{-1} \\
\tfrac{1}{2}Cl_{2,g} &= Cl_g & \Delta H_3 &= 121 \text{ kJ mole}^{-1} \\
Cl_g + e^- &= Cl^-_g & \Delta H_4 &= -369 \text{ kJ mole}^{-1} \\
Ca^+_g + Cl^-_g &= Ca^+Cl^-_s & \Delta H_5 &- 865 \text{ kJ mole}^{-1} \text{ (estimated lattice energy*)}
\end{aligned}$$

Adding,

$$Ca_s + \tfrac{1}{2}Cl_{2,g} = Ca^+Cl^-_s \quad \Delta H_f = -349 \text{ kJ mole}^{-1}$$

This compares with a heat of formation of -796 kJ mole^{-1} for $CaCl_2$.

* The lattice energy for Ca^+Cl^- is estimated as the mean of the values for Na^+Cl^- and K^+Cl^-. This estimate seems reasonable as the atomic size of calcium (and therefore the probable size of the ion Ca^+) falls between that of sodium and potassium (see also p. 122).

Now, if we express these heats of formation as thermochemical equations:

$$Ca_s + \tfrac{1}{2}Cl_2 = CaCl_s \quad \Delta H_f = -349 \text{ kJ mole}^{-1} \quad (1)$$
$$Ca_s + Cl_2 = CaCl_{2,s} \quad \Delta H_f = -796 \text{ kJ mole}^{-1} \quad (2)$$

Multiplying (1) by two and subtracting the result from (2) we get

$$2CaCl_s = Ca_s + CaCl_{2,s} \quad \Delta H = -98 \text{ kJ mole}^{-1} \quad (3)$$

This means that the compound CaCl is stable with respect to its elements (i.e. energy is released when it is formed from its elements) but is unstable with respect to $CaCl_2$ as yet more energy is released when reaction 3 takes place.

The calculation of electron affinities Provided a reasonable estimate can be made for the lattice energy of an ionic compound, the electron affinity of a non-metal may be calculated from a Born–Haber cycle. This is illustrated in Example 42.

Example 42 The lattice energy of sodium fluoride is estimated by the method given on p. 123 to be 903 kJ mole^{-1}, while the heat of atomization of sodium is 109 kJ mole^{-1}. The first ionization energy of sodium is 494 kJ mole^{-1}. The bond dissociation energy of fluorine is 160 kJ mole^{-1} and the heat of formation of sodium fluoride is $\Delta H_f = -572$ kJ mole^{-1}. Estimate the electron affinity of fluorine.

Solution Setting up the Born–Haber cycle:

$$Na_s = Na_g \quad \Delta H_1 = 109 \text{ kJ}$$
$$Na_g = Na_g^+ + e^- \quad \Delta H_2 = 494 \text{ kJ}$$
$$\tfrac{1}{2}F_{2g} = F_g \quad \Delta H_3 = 80 \text{ kJ}$$
$$F_g + e^- = F_g^- \quad \Delta H_4 = ?$$
$$Na_g^+ + F_g^- = Na^+F_s^- \quad \Delta H_5 = -903 \text{ kJ}$$

Also

$$Na_s + \tfrac{1}{2}F_{2,g} = Na^+F_s^- \quad \Delta H_f = -572 \text{ kJ}$$

Equating the two parts of the cycle,

$$\Delta H_f = \Delta H_1 + \Delta H_2 + \Delta H_3 + \Delta H_4 + \Delta H_5$$
$$-572 = -220 + \Delta H_4$$
$$\Delta H_4 = -352 \text{ kJ mole}^{-1}$$

which is the electron affinity of fluorine.

Factors governing the size of lattice energy Lattice energy is a measure of the force holding the ions in their positions in a crystal lattice. Now, the force between two charges q_1 and q_2 situated a distance r apart is

$$\text{Force} = \frac{q_1 q_2}{r^2}$$

while the energy corresponding to this force is given by

$$E = \frac{q_1 q_2}{r}$$

Thus, two factors are mainly responsible for the magnitude of lattice energy:

(a) The size of the charge on the ions—as the charge increases so the lattice energy increases. This can be seen from the following values:

Salt	NaCl	CaCl$_2$	CaO
Lattice energy kJ mole^{-1}	918	2204	3448

(b) The distance apart of the ions. This is governed by the size of the ions and the manner in which they are packed in the crystal lattice. If the distance between the ions is regarded as being the sum of the anion and cation radii, then an increase in lattice energy follows a decrease in ionic sizes. This can be seen from the values set out below:

Salt	$r_{(cation)}$	$r_{(anion)}$	$r_+ + r_-$	Lattice energy
RbI	1·5 Å	2·2 Å	3·7 Å	624 kJ mole^{-1}
KBr	1·3	2·0	3·3	679
NaCl	0·9	1·8	2·7	779
LiF	0·6	1·4	2·0	1031

Calculation of lattice energies The energy obtaining between two charges is given by

$$E = \frac{q_1 q_2}{r} = \frac{z_1 e \times z_2 e}{r}$$

where z_1 and z_2 represent the integral charges on the cation and anion, and e is the electronic charge. The sum of the cation and anion radii is equal to r. Since ions of opposite charge attract, the energy relating to an attractive force will bear a negative sign (z_1 is positive, z_2 is negative). To prevent the lattice from collapsing, it is assumed that certain repulsive forces operate within the lattice, and these forces become large as r becomes very small. To allow for this, an extra term is introduced into the energy equation.

$$E = \frac{z_1 z_2 e^2}{r} - \text{repulsion term}$$

For one mole of ions present in a crystal, the equation becomes

$$E = \frac{N z_1 z_2 e^2 A}{r} - \text{repulsion term}$$

where A is a factor (called the *Madelung* constant) which allows for the type of packing present in the crystal structure, and N is Avogadro's number. The equation proposed by M. Born (1918) and thus known as the *Born equation* includes the repulsion energy in terms of r, and is

$$E = \frac{N e^2 z_1 z_2 A}{r}\left(1 - \frac{1}{3r}\right)$$

If r is in ångstroms, the value of E in kJ mole^{-1} is given by

$$E = \frac{1383 z_1 z_2 A}{r}\left(1 - \frac{1}{3r}\right)$$

Example 43 Calculate the lattice energy of sodium fluoride which has a rock salt structure (Madelung constant 1·75), and for which the cation and anion radii are 0·93 and 1·36 Å respectively.

Solution Using the Born equation,

$$E = \frac{1383 \times (+1) \times (-1) \times 1\cdot 75}{2\cdot 29}\left(1 - \frac{1}{6\cdot 87}\right)$$

$$= -903 \text{ kJ mole}^{-1}$$

Theoretical and experimental lattice energies The Born equation used for the calculation of lattice energies is based on an ideal ionic crystal in which the ions are point charges, located at a fixed distance

Salt	Observed lattice energy kJ mole^{-1}	Lattice energy calculated from Born equation kJ mole^{-1}	Difference	Difference as a percentage of observed value
NaF	918	903	15	1·6
NaCl	779	769	10	1·3
KCl	708	693	15	2·1
MgF$_2$	2925	2920	5	0·2
CaF$_2$	2635	2595	40	1·5
AgCl	901	850	51	5·7
AgBr	888	825	63	7·2
AgI	884	796	88	10·0
CdI$_2$	2359	1986	373	15·8

apart, and the linkages are fully ionic. The fact that many of the calculated values of lattice energy agree with those obtained experimentally by means of the Born–Haber cycle is a valuable piece of evidence in favour of the model of ionic bonding on which the Born equation depends.

The discrepancy between the observed and calculated values for the silver halides agrees with the increasing polarizability* of the anion from chloride to iodide. This may be taken as an indication of some covalent character being present in the bonding, and the large deviation for cadmium iodide shows that electron transfer is far from complete in this instance.

Covalent Bonding

In this type of bonding, the electrostatic forces operating between neighbouring atoms result in an intermingling of valence electrons, and our picture of a covalent bond is based on a shared pair of electrons forming the link between two atoms. To break apart atoms linked in this way, energy, described in general terms as bond energy, is required, while on the formation of covalent links between atoms, bond energy is released. For example, the reaction

$$H_2 + Cl_2 = 2HCl \quad (\Delta H = 2 \times (-91 \cdot 5) \text{ kJ mole}^{-1})$$

can be regarded as the sequence

$$H_2 + Cl_2 \rightarrow \underset{\text{atoms}}{2H} + \underset{\text{atoms}}{2Cl}$$
$$\downarrow$$
$$2HCl \text{ molecules}$$

Now 436 kJ of energy are required to produce 2 moles of hydrogen atoms from one mole of hydrogen gas, similarly, 243 kJ are needed for chlorine. The total energy released on combination of 2 moles of hydrogen atoms and 2 moles of chlorine atoms is 862 kJ. Thus, the total energy transfer is

$$436 + 243 - 862 = -183 \text{ kJ}$$

The energy released per mole of hydrogen chloride is 91·5 kJ which

---

* Polarizability refers to the distortion of the electron screen surrounding an ion or molecule. The peripheral electrons of a large anion are not strongly bound to the nucleus, so that, on the approach of a small or highly charged cation, they may be pulled towards the cation. This distortion (polarization) may be so great as to result in some degree of interaction between the outer electrons in the anion and the cation. This interaction is referred to as the *covalent character* of an ionic bond.

equals the *heat of formation* of hydrogen chloride. The heat of formation represents the balance between the large amounts of energy required to break the bonds linking the reactant atoms and that released on formation of the product molecules. In consequence, the heat of formation does not give a true indication of the strength of the bonds involved.

Determination of bond energies The bond energy of a diatomic molecule is defined as the energy required to break apart one mole of molecules into atoms. (Such values are obtained spectroscopically or from measurements of dissociation constants at high temperatures.) In the case of polyatomic molecules, the bond energy may be calculated using Hess's law as illustrated in Example 44.

Example 44 The heat of formation of water (as steam) is $\Delta H_f = -243$ kJ mole^{-1}. The bond dissociation energies of hydrogen and oxygen are 436 kJ mole^{-1} and 494 kJ mole^{-1} respectively. Calculate the average bond energy of the H—O bond.

Solution Writing thermochemical equations for the given data:

$$H_{2,g} = 2H_g \qquad \Delta H_1 = 436 \text{ kJ}$$
$$\tfrac{1}{2}O_{2,g} = O_g \qquad \Delta H_2 = 494/2 \text{ kJ}$$
$$2H_g + O_g = H_2O_g \qquad \Delta H_3 = ?$$

Also

$$H_{2,g} + \tfrac{1}{2}O_{2,g} = H_2O_g \qquad \Delta H_f = -243 \text{ kJ}$$

Equating the two routes,

$$\Delta H_1 + \Delta H_2 + \Delta H_3 = \Delta H_f$$
$$436 + 247 + \Delta H_3 = -243$$
$$\Delta H_3 = -926 \text{ kJ}$$

Now ΔH_3 refers to the *production* of two moles of H—O bonds. Hence, the average bond energy (energy required to break the bond) of the H—O link is

$$926/2 = 463 \text{ kJ mole}^{-1}$$

In Example 44, the final result is the *average* bond energy as the two H—O links are taken to be equal in strength. The average bond energy is denoted by writing

$$E_{H-O} = 463 \text{ kJ mole}^{-1}$$

However, this assumption is not true, as the successive stages

$$2H + O \rightarrow H + O{-}H \rightarrow H{-}O{-}H$$

do not release equal quantities of energy. The term *bond dissociation energy* is used when the energy required to break a specific bond (i.e. the C—Cl bond in CH_3Cl) is to be considered. Bond dissociation energies are denoted by writing (for example) $D_{C-Cl;CH_3Cl}$. Bond energy values given for singly bonded diatomic molecules are, in fact, bond dissociation energies.

Example 45 The heat of combustion of methane is $\Delta H = -805$ kJ mole^{-1} while the heats of formation of water and carbon dioxide are -243 kJ mole^{-1} and -394 kJ mole^{-1} respectively. If the heat of atomization of graphite is 716 kJ mole^{-1} and the bond dissociation energy of hydrogen is 433 kJ mole^{-1}, calculate the average C—H bond energy in methane.

Solution Writing thermochemical equations for the data given:

$$CH_4 + 2O_2 = CO_2 + 2H_2O \quad \Delta H = -805 \text{ kJ}$$

$C_s = C_g$ (atoms)	$\Delta H_1 = 716$ kJ	(1)
$2H_{2,g} = 4H_g$	$\Delta H_2 = 2 \times 433$ kJ	(2)
$C_s + O_{2,g} = CO_{2,g}$	$\Delta H_3 = -394$ kJ	(3)
$2H_{2,g} + O_{2,g} = 2H_2O_g$	$\Delta H_4 = -2 \times 243$ kJ	(4)
$C_g + 4H_g = CH_4$	$\Delta H_5 = ?$	(5)

Adding equations (3) and (4) and subtracting equations (1), (2) and (5) we get

$$CH_4 + 2O_2 = CO_2 + 2H_2O$$

and for this reaction,

$$\Delta H = \Delta H_3 + \Delta H_4 - \Delta H_1 - \Delta H_2 - \Delta H_5$$
$$= -2462 - \Delta H_5$$

This quantity, ΔH, can be equated to the observed heat of combustion.

$$-2462 - \Delta H_5 = -805$$
$$\Delta H_5 = -1657 \text{ kJ}$$

which is the energy released on the formation of four C—H bonds in methane. Hence

$$E_{C-H} = 1657/4 = 414 \text{ kJ mole}^{-1}$$

By means of a similar calculation, it may be shown that for ethane,

$$2C_{g(atoms)} + 6H_{g(atoms)} = C_2H_6 \quad \Delta H = -2826 \text{ kJ mole}^{-1}$$

Now ethane contains six C—H bonds, average bond strength 414 kJ mole^{-1} (assuming the strength is the same as in methane) and these account for 2484 kJ of the total bond energy. The difference $(2826-2484 = 342 \text{ kJ mole}^{-1})$ is taken to correspond with the energy of the C—C single bond found in ethane.

In this calculation, we have assumed that the total bond energy of a molecule is the sum of the individual (average) bond energies. Therefore bond energies are regarded as *additive*, and we can estimate the heat of formation of a substance by adding up the energy contributed by each bond in the molecule together with the heats of atomization of the constituent elements.

Example 46 If $E_{\text{C—C}}$ is 344 kJ mole^{-1} and $E_{\text{C—H}}$ is 415 kJ mole^{-1}, calculate the heat of formation of propane. The heats of atomization of carbon and hydrogen are 716 kJ mole^{-1} and 433 kJ mole^{-1} respectively.

Solution The heat of formation is the sum of the heats of atomization and bond energies. For propane, the heats of atomization are

$$3C_s = 3C_g \quad \Delta H = 3 \times 716 = 2148 \text{ kJ}$$
$$4H_{2,g} = 8H_g \quad \Delta H = 4 \times 433 = 1732 \text{ kJ}$$

The bond energies are

$$2E_{\text{C—C}} = 2 \times -344 = -688 \text{ kJ}$$
$$8E_{\text{C—H}} = 8 \times -415 = -3320 \text{ kJ}$$

Adding,

$$3C + 4H_2 = C_3H_8 \quad \Delta H_f = 2148 + 1732 - 688 - 3320$$
$$= -128 \text{ kJ mole}^{-1}$$

This compares with the observed value of -124 kJ mole^{-1}. Some heats of atomization and bond energies are given below, a more complete list appears on page 244 at the end of the book.

Heats of Atomization

$$C_s = C_g \quad \Delta H = 716 \text{ kJ mole}^{-1}$$
$$H_{2,g} = 2H_g \quad \Delta H = 433 \text{ kJ mole}^{-1}$$

Bond Energies (kJ mole^{-1})

C—C	344	C—H	415	O—H	461
C=C	611	C—O	352	C—Cl	327
C≡C	836	C=O	742	H—Cl	431

The proposal that bond energies should be additive (which means that the energy of a chemical bond remains unchanged irrespective of its environment) applies only to molecules of the same type and where no complicated structural features occur. The alkanes form a good example of this. Conversely, a heat of formation calculated from bond energies which differs significantly from the observed value is an indication that the proposed structure for the molecule under consideration is inadequate.

Example 47 Using the data given in the text, calculate the heats of formation of cyclohexane and benzene.

Solution The proposed structure for cyclohexane is

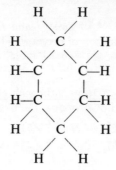

The heats of atomization are

$$6C_s = 6C_g \quad \Delta H = 4296 \text{ kJ}$$
$$6H_{2,g} = 12H_g \quad \Delta H = 2598 \text{ kJ}$$

The bond energies are

$$6E_{C-C} = 6 \times -344 = -2064 \text{ kJ}$$
$$12E_{C-H} = 12 \times -415 = -4980 \text{ kJ}$$

Adding,
$$\Delta H_f = -150 \text{ kJ mole}^{-1}$$

(This compares with a value of -151 kJ mole^{-1} found experimentally.)

Taking the structure for benzene as

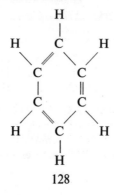

Heats of atomization:

$$6C_s = 6C_g \quad \Delta H = 4296 \text{ kJ}$$
$$3H_{2,g} = 6H_g \quad \Delta H = 1299 \text{ kJ}$$

Bond energies:

$$3E_{C-C} = -1032 \text{ kJ}$$
$$3E_{C=C} = -1833 \text{ kJ}$$
$$6E_{C-H} = -2490 \text{ kJ}$$

Adding, $\Delta H_f = +240$ kJ mole^{-1} which compares with an experimental value of 84 kJ mole^{-1}. In view of this large discrepancy, the assumed structure is clearly inadequate. This is explained in terms of delocalization of electrons (Part 1, p. 34) and a similar conclusion was arrived at on page 115 in connection with heats of hydrogenation.

Bond Energies and Reactivity

In the study of chemistry, we are continually seeking to explain why things happen in the way they do. Frequently, the immediate answer is given in terms of theoretical concepts—bond energies being a typical example. However, in giving these explanations, we must be careful not to quote values out of context, or ignore other changes that are relevant to the problem, which may be taking place. For example, nitrogen (assumed structure N≡N) is an unreactive molecule, while ethyne (H—C≡C—H) is highly reactive. Yet, the triple bond energy in nitrogen is 947 kJ mole^{-1} while that in ethyne is about 900 kJ mole^{-1}. Obviously, the bond energies are too close to explain the difference in reactivities, and we need to consider the reaction as a whole. Taking hydrogenation as an example:

$$N_2 + 3H_2 = 2NH_3$$

Energy needed to break bonds: 947 1308
$$\underbrace{}_{2255}$$

Energy released on bond formation: 2×1170
$$\underbrace{}_{2340}$$

Hence, the bonds in the reactants are being replaced by bonds which are very slightly stronger in the product molecules.

In the case of ethyne:

$$H-C \equiv C-H + 2H_2 = C_2H_6$$

Energy needed to break bonds: $\underbrace{900 + 2 \times 415 \quad 872}_{2602}$

Energy released on bond formation: $\underbrace{2 \times 344 + 6 \times 415}_{3178}$

In this case, the bonds in the reactants are replaced by much stronger bonds in the products. (Entropy changes—see p. 135—have been neglected.)

Example 48 The heats of formation of PCl_3 and PH_3 are -306 kJ mole^{-1} and $+8$ kJ mole^{-1} respectively, and the heats of atomization of phosphorus, chlorine and hydrogen are given by:

$$P_s = P_g \quad \Delta H = 314 \text{ kJ mole}^{-1}$$
$$Cl_{2,g} = 2Cl_g \quad \Delta H = 242 \text{ kJ mole}^{-1}$$
$$H_{2,g} = 2H_g \quad \Delta H = 433 \text{ kJ mole}^{-1}$$

Calculate E_{P-Cl} and E_{P-H} and comment on the statement that 'The instability of PH_3 compared with PCl_3 is a reflection of the low P—H bond energy'.

Solution For phosphorus trichloride, the thermochemical equations are:

$$P_s = P_g \quad \Delta H_1 = 314 \text{ kJ}$$
$$\tfrac{3}{2}Cl_{2,g} = 3Cl_g \quad \Delta H_2 = 363 \text{ kJ}$$
$$P_g + 3Cl_g = PCl_3 \quad \Delta H_3 = ?$$

Also,

$$P_s + \tfrac{3}{2}Cl_{2,g} = PCl_3 \quad \Delta H_f = -306 \text{ kJ}$$

Equating the two routes leading to the formation of PCl_3,

$$\Delta H_f = \Delta H_1 + \Delta H_2 + \Delta H_3$$
$$-306 = 314 + 363 + \Delta H_3$$
$$\Delta H_3 = -983$$

This is the energy released on the formation of three P—Cl bonds, hence,

$$E_{P-Cl} = \frac{983}{3} = 328 \text{ kJ mole}^{-1}$$

Similarly, for phosphine,

$$P_s = P_g \quad \Delta H_1 = 314 \text{ kJ}$$
$$\tfrac{3}{2}H_{2,g} = 3H_g \quad \Delta H_2 = 649$$
$$P_g + 3H_g = PH_3 \quad \Delta H_3 = ?$$

Also,
$$P_s + \tfrac{3}{2}H_{2,g} = PH_3 \quad \Delta H_f = 8 \text{ kJ}$$

From which
$$\Delta H_3 = -958 \text{ kJ}$$

$$E_{P-H} = \frac{958}{3} = 319 \text{ kJ}$$

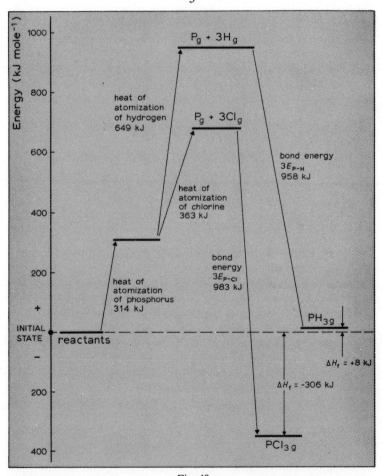

Fig. 48

Thus, the bond energies of the single bonds P—Cl and P—H are in fact very close together, and the statement is not justified. If the formations of the two compounds are considered diagrammatically (Figure 48), we can see that the large single bond energy of hydrogen compared with that of chlorine is the factor which leads to the large difference in the heats of formation.

Bond energies and electronegativity In a singly bonded diatomic molecule, the two atoms are linked by means of a pair of electrons located between the bonded atoms. If the two atoms are identical (as in a fluorine molecule, F_2) we may assume that the bonding electrons are shared equally between the two atoms. For molecules in which the atoms are different (such as hydrogen fluoride, HF) the bonding electrons will most probably not be equally shared between the two atoms, and this has an important effect on the bond energy. Suppose we make an attempt to estimate the strength of the single covalent bond in hydrogen iodide using the following model. Let us regard the electrons (shown by crosses in Figure 49) linking the two

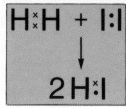

Fig. 49

hydrogen atoms as a concentrated 'glue' with an adhesive power of (say) 434 units. Similarly, the dotted electrons linking the two iodine atoms are a more dilute 'glue' with an adhesive power of (say) 150 units. Now let us mix equal portions of concentrated and dilute glue (i.e. one dot and one cross electron) and bond together a hydrogen and an iodine atom with this less powerful adhesive. We would expect the strength of this union to be the exact average (i.e. 292 units) of the strength of the initial hydrogen and iodine bonds. Using the fact that D_{H-H} is 434 kJ mole^{-1} and D_{I-I} is 150 kJ mole^{-1}, we would estimate the value of E_{H-I} to be

$$\begin{aligned} E_{H-I} &= (\tfrac{1}{2}D_{H-H} + \tfrac{1}{2}D_{I-I}) \\ &= 217 + 75 \\ &= 292 \text{ kJ mole}^{-1} \end{aligned}$$

The experimental value is 297 kJ mole^{-1} which is close to our estimate. However, in most cases, the estimated values fall short of the observed single bond energies as shown in the following list. (All energy values in kJ mole^{-1}.)

Bond	Estimated (or pure covalent) bond energy	Observed bond energy	$\Delta = Obs. - Est.$
H–I	292	297	5
H–Br	314	364	50
H–Cl	339	431	92
H–F	297	566	269
C–H	390	414	24
C–O	243	352	109
C–F	251	440	189

The increase over the expected pure covalent bond energy is related to the fact that, in most bonds linking unlike atoms, the sharing of the bonding electrons is unequal. This results in a slight excess of negative charge (denoted by $\delta-$ in Figure 50) residing close to one atom, and a slight excess of positive charge ($\delta+$) near the other. Such a bond is said to be polar, i.e. it has a degree of ionic character, and this adds to the covalent bond energy as estimated above. Atom B, Figure 50, attracts the electron pair most strongly and it

Fig. 50

is thus said to be more electronegative than A. (See Part 1, p. 44). Pauling used the Δ values* to draw up his scale of electronegativities, and the electronegativity concept has useful applications in both inorganic and organic chemistry. Any polar bond has a corresponding *dipole moment* μ, given by the relation $\mu = zd$ where z is the size of the charge and d is the distance between the charges. A polar bond is indicated by writing an arrow over the bond pointing to the more electronegative element, e.g. A$\overset{\rightarrow}{-}$B. The dipole moment of a molecule is the vector sum of the bond moments, hence for symmetrical molecules the dipole moment is zero, although the bonds are polar (Figure 51). It is obvious that dipole moment determinations are of great importance in the elucidation of molecular structures. Dipole moments are measured in Debye units (D). A polar bond differs from a purely covalent bond in that:

* The average bond energies used by Pauling in calculating Δ values were geometric and not arithmetic means. However, the principle is the same, and arithmetic means have been used here so as not to complicate the calculations and obscure the theme of the discussion.

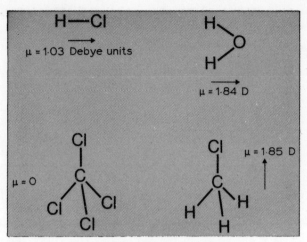

Fig. 51. Dipole moments of molecules

(a) The bond is shorter than the sum of the two covalent radii. This contraction is related to the electronegativity difference between the two atoms.

(b) The bond is subject to attack by ions or other charged groups. This factor is of special significance when studying the hydrolysis of covalent compounds (Part 1, p. 55) and organic reaction mechanisms as outlined in Part 2.

Factors Governing a Chemical Reaction

Before a chemical reaction can take place, the reacting molecules must be brought together by way of a collision. We can draw up a simple analogy between chemical reaction and trade which we shall follow for some way before re-stating our ideas in physical chemical terms. In business, before a sale (i.e. reaction) can take place, the vendor, a shopkeeper say, and the purchaser have to make contact. Money (energy) is exchanged and the shopkeeper continues to trade provided he makes a profit. Similarly, a reaction can continue provided energy is released.

$$\text{Reactants} \rightarrow \text{Products} + \text{Energy released (profit)}$$

The energy exchange is designated ΔH.

Our shopkeeper may use his profit in two ways:

(a) he may plough part of it back into his business by way of renewing and extending his stock;

(b) he may feel free to spend some of his profit in any way he wishes.

Thus we can say

Profit = free money + money retained as stock etc.
ΔH = free energy + energy retained by the system

By increasing and extending his stock, our shopkeeper gradually fills up shelves and space not previously occupied, until the shop becomes cluttered with a disorderly array of articles (which may eventually produce a chaotic situation). By increasing his stock, the shopkeeper *increases the amount of disorder* within the shop. So we can continue the analogy:

ΔH = ΔG + energy retained to increase the disorder of
total energy   free   the system
released   energy

The disorder of a system is called the *entropy* of the system (symbol S) while ΔS is the increase in the disorder following a reaction. The units of entropy are joules per degree Kelvin per mole. However, both ΔH and ΔG are expressed in joules, so that to express all the quantities in the same units, ΔS must be multiplied by the Kelvin temperature, to give

$$\Delta H = \Delta G + T\Delta S \qquad (1)$$

We said earlier that our shopkeeper remains in business provided he continues to make a profit. What we really meant was that the shopkeeper remains in business provided his profit allows him to renew his stock and leaves him enough free income to pay for his standard of living. That is, it is the establishment of a sufficient free income (ΔG) which is the deciding factor.

Thus, we re-write equation (1) as

$$\Delta G = \Delta H - T\Delta S$$

and the question of whether a reaction may proceed or not hinges on the value of ΔG. Now a release of energy is always indicated by a negative sign. Hence, for the reaction

$$\text{Reactants} \rightarrow \text{Products}$$

to have the ability to proceed from left to right, the corresponding *value of ΔG must be negative.*

So far, we have looked at a reaction from the point of view of energy changes—this is the thermodynamic aspect. There is also a kinetic aspect. Again, to use the shopkeeping analogy, it could well be that many people are eager to buy goods from our shopkeeper which are in short supply and are temporarily out of stock. It may also happen that these items are imported, and the mechanics of ordering, obtaining Government trade sanctions, shipping and transportation, clearance through the Customs, etc., are so involved that the items come through at a very slow rate. Although the shopkeeper makes a profit on these sales, if the mechanism of obtaining the goods is complex his turnover may be so slow that his business fails. The same is true of chemical reactions, which may be thermodynamically feasible (i.e. ΔG is negative), but take place by such a complex mechanism that they proceed at an infinitesimally slow rate. This second aspect, the kinetic factor, is discussed further on p. 162.

Free energy and entropy Every reaction is accompanied by:
  (a) a heat (energy) change, stated in terms of ΔH
  (b) a redistribution of matter, given in terms of ΔS,

and we have seen from our shopkeeping analogy that it is a combination of these two factors through the equation

$$\Delta G = \Delta H - T\Delta S$$

which decides whether a reaction is thermodynamically feasible or not at a constant temperature and pressure.

It is our experience that changes take place naturally in a direction corresponding to a loss in energy. Water flows downhill, losing potential energy (which is converted into kinetic energy and finally dissipated as heat), a coiled spring shoots out losing its energy of compression. Considering the redistribution of matter, we know from experience that any system in random movement becomes more mixed up or disorderly as time passes (e.g. the mixing of separate layers of salt and sand when tumbled in a sealed tube). Now order is linked with both physical state and molecular structure. Liquids are more ordered than gases, polyatomic molecules are more ordered than the collection of small molecules from which they may be produced. Substituting the word entropy to indicate degree of disorder, we can say that the entropy of a liquid is greater than that of a solid, or the entropy of a solution is greater than that of the separated solute and solvent, while the entropy of a hot body is greater than that of a cold one. Heat energy flows spontaneously from a hotter to a cooler body, raising the level of disorder in the cooler body by ΔS and causing a rise in temperature or change of state. At zero Kelvin, a purely crystalline material is a completely ordered structure and its entropy is zero. If q joules of heat energy are given to this crystal, the temperature rises to T K and there is an increase in entropy ΔS. It can be shown that

$$q = T\Delta S$$

which re-arranges to give
$$\Delta S = q/T$$

(Note $\Delta S = S_{\text{final state}} - S_{\text{initial state}}$.)

Example 49 Calculate the change in entropy when one mole of water is vaporized at 100°C. Latent heat of vaporization of water is 37·8 kJ mole^{-1}.

Solution Using the equation $\Delta S = q/T$,
$$q = 37\cdot 8 \text{ kJ}, \quad T = 373 \text{ K}$$

Thus
$$\Delta S = \frac{37\cdot 8}{373} = 0\cdot 101 \text{ kJ mole}^{-1}$$
$$= 101 \text{ J mole}^{-1}$$

Solids have a high degree of order, liquids less so, while in gases molecules are in random motion. Hence, we may conclude that the entropy of a gas is much larger than that of a liquid which is in turn greater than that of a solid, i.e.

$$S_{\text{gas}} \gg S_{\text{liq}} > S_{\text{solid}}$$

(For example, the entropies of NaF, H_2O liquid and steam are 51·6, 70·0 and 192·5 J deg^{-1} mole^{-1} respectively at 298 K.) Thus, any chemical reaction in which a greater number of moles of gas appear on the product side than on the reactant side is accompanied by a large increase in entropy.

For example, the entropy change accompanying the reaction

$$C + O_2 = CO_2$$

is very small ($\Delta S = 3$ J mole^{-1} deg^{-1}) since there is one mole of gas on both sides of the equation. But the reaction

$$2C + O_2 = 2CO$$

is accompanied by a large increase in entropy ($\Delta S = 179$ J mole^{-1} deg^{-1}) due to the presence of an extra mole of gaseous product.

In deciding whether a reaction is theoretically possible, we have to evaluate the free energy change (ΔG) for the reaction, using the equation

$$\Delta G = \Delta H - T\Delta S$$

It is the *sign* of ΔG which is critical. If ΔG is negative, the reaction is possible, if ΔG is positive, the reaction is considered unlikely, while ΔG values of zero (or close to zero) shows that the reaction comes to equilibrium.*

Without knowing an actual value, it is often possible to estimate the sign of ΔG by examining the two quantities on the right of the above equation.

Considering the enthalpy factor, ΔH, exothermic reactions are more likely to take place than endothermic reactions. The sign of ΔH may be estimated with reference to bond strengths. If the sum of the bond strengths of the product molecules exceeds that of the reactants, then the reaction will be exothermic. In addition, at low temperatures (T small) the value of $T\Delta S$ tends to become insignificant compared with ΔH. Thus, at low temperatures, the enthalpy change determines the favoured direction of reaction.

In the entropy term, T is always positive, so that the sign of this term depends on ΔS. An *increase* in entropy (ΔS is positive) gives rise to a negative contribution to ΔG. Therefore, favoured reactions are those which are accompanied by an increase in entropy. At high temperatures, the $T\Delta S$ term becomes large and the course of the reaction at high temperatures is governed by the entropy factor. It follows that, at very high temperatures, all systems react to give an increase in entropy, i.e. large molecules break down to give smaller molecules which in turn dissociate into atoms.

Estimation of free energy changes for selected reactions

1. *Exothermic reactions accompanied by an increase in entropy* The thermal decomposition of ammonium dichromate or ammonium nitrite are examples.

$$(NH_4)_2 Cr_2 O_7 = N_2 + Cr_2 O_3 + 4H_2 O$$

We would estimate ΔH to be negative as both N_2 and H_2O contain strong bonds, while five moles of gaseous product are formed during this reaction. This indicates a very large increase in entropy.

$$NH_4 NO_2 = N_2 + 2H_2 O$$

Again, we would expect ΔH to be negative and ΔS positive for

* In using ΔG values in this way, all reactions are treated as equilibrium reactions, for which it can be shown that

$$\Delta G = -RT \log_e K$$

where K is the equilibrium constant. Hence, a large negative value for ΔG means that the equilibrium lies over to the extreme right (i.e. reactants give products) while a large positive ΔG value shows the equilibrium position lies to the far left of the reaction (products give reactants).

similar reasons. In both reactions, we would predict large negative values for ΔG.

2. *Endothermic reactions accompanied by a decrease in entropy* Such reactions are, of course, the exact converse of those discussed under the previous heading.
3. *Exothermic reactions accompanied by a decrease in entropy* Typical examples are:

$$H_2 + \tfrac{1}{2}O_2 = H_2O$$
$$N_2 + 3H_2 \rightleftharpoons 2NH_3$$
$$2SO_2 + O_2 \rightleftharpoons 2SO_3$$

In all these reactions, the products occupy a smaller gaseous volume than the reactants. For the first reaction, the decrease in entropy is small, but ΔH has a large negative value, so that ΔG has a high negative value. We would predict that this reaction is probable at lower temperatures, but becomes less favoured at higher temperatures. An equilibrium is set up in the other two reactions, as the entropy decrease is larger in these two cases. Again, ΔG has an overall negative value at lower temperatures, but the reactions become less probable as the temperature rises, since the $T\Delta S$ term increases.

4. *Endothermic reactions accompanied by an increase in entropy* Typical examples are the vaporization of a liquid, the dissolution of a solid and reactions such as:

$$NH_4Cl_{(s)} \rightleftharpoons NH_3 + HCl$$
$$PCl_5 \rightleftharpoons PCl_3 + Cl_2$$
$$N_2O_4 \rightleftharpoons 2NO_2$$
$$CaCO_3 \rightleftharpoons CaO + CO_2$$

In all these cases, a more disorderly system is produced, although energy is absorbed in the forward reaction. Thus, we would predict that although the equilibrium may lie towards the left at low temperatures, the formation of the products becomes increasingly favoured at high temperatures. Thus, a vapour condenses on cooling and evaporates on heating, while solids such as ammonium nitrate or potassium iodide, which are freely soluble, dissolve in water with the absorption of heat (the solution cools) to form a less ordered system.

Chemical Reduction of Metal Oxides

Carbon is commonly used as a reducing agent in the process of extracting metals from their oxide ores. In the reduction reaction, carbon may be oxidized to either carbon dioxide or carbon monoxide.

$$C_s + O_{2,g} = CO_{2,g} \quad (\Delta G \text{ at } 25°C = -390 \text{ kJ mole}^{-1})$$

In this case, there is little change in entropy, as one mole of gas is found on each side of the equation. But for carbon monoxide, an increase in entropy occurs.

$$2C_s + O_{2,g} = 2CO \quad (\Delta G \text{ at } 25°C = -280 \text{ kJ mole}^{-1})$$

(Note that the ΔG values quoted in this connection are given per mole of oxygen taken up in the reaction, since, in the reduction of metal oxides, we wish to compare the acceptance of equal amounts of oxygen by the reducing agent.)

Now the variation of ΔH and ΔS with temperature is small, and they may be regarded as constants. Hence, a graph of ΔG against T will take the form of a straight line with slope $-\Delta S$.

$$\Delta G = -\Delta S T + \Delta H$$

(compare with $y = -mx + c$)

This is shown in Figure 52, which is known as an *Ellingham diagram*. Since the value for ΔS is almost zero for carbon dioxide, the straight line runs almost parallel to the temperature axis. However, for carbon monoxide, which shows an increase in entropy on formation from its elements, ΔS is positive and ΔG for the reaction becomes increasingly negative as the temperature increases.

Taking zinc oxide as a typical example, we can see from the equation

$$2Zn_s + O_2 = 2ZnO_s$$

that there is a decrease in entropy accompanying the formation of the metal oxide. Thus ΔS is negative and ΔG, which is -628 kJ at 25°C, increases (i.e. becomes less negative) as the temperature rises. (The straight line for zinc oxide shown in Figure 52 has two breaks in it corresponding to the melting point and boiling point of zinc. At each point, the slope increases, since the entropy decrease for the reaction

$$2Zn + O_2 = 2ZnO$$

is even greater when liquid zinc or zinc vapour forms the starting point.) There is a temperature when the free energy change for the two reactions,

$$2C + O_2 = 2CO$$
$$2Zn + O_2 = 2ZnO$$

becomes equal, and above this temperature it is possible to reduce zinc oxide with carbon, the products being zinc vapour and carbon monoxide. For example, at 1000°C,

$$2C + O_2 = 2CO \quad \Delta G = -440 \text{ kJ}$$
$$2Zn + O_2 = 2ZnO \quad \Delta G = -378 \text{ kJ}$$

Subtracting and halving

$$C + ZnO = CO + Zn: \quad \Delta G = -31 \text{ kJ mole}^{-1}$$

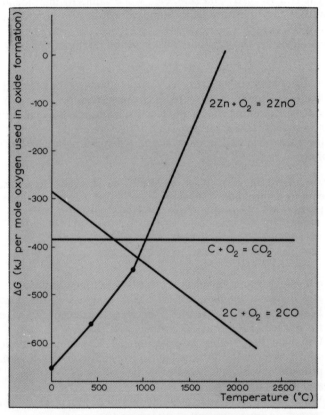

Fig. 52. Ellingham diagram

A further account of the use of free energy data in the form of an Ellingham diagram is given in Part 1, p. 69.

Since ΔH and ΔS remain fairly constant over a range of temperature, the fact that $T\Delta S$, and not ΔS, is the term contributing to the ΔG value means that temperature has a significant effect on the course of a reaction. This is particularly true in the case of endothermic reactions accompanied by an increase in entropy, which may change direction as the temperature rises.

Standard states It has already been mentioned on p. 110 that thermochemical quantities are given values appropriate to 298 K and one atmosphere pressure, with the reactants and products in their normal (or otherwise specifically defined) states. Such values are designated ΔH^0, ΔG^0 and ΔS^0, etc., while values of thermochemical quantities not appropriate to standard states are designated ΔH, ΔH or ΔS, etc.

In conclusion, the possibility of a reaction taking place is governed by:

1. (a) the change in enthalpy (ΔH) accompanying the reaction,
   (b) the change in entropy (ΔS) of the system,
   and the combined effects of these two factors is shown in terms of a free energy (ΔG) change;
2. the mechanism of the reaction and the speed with which it takes place.

It is likely, therefore, that we shall find cases where a reaction appears to be probable (ΔG is negative) but proceeds so slowly that the formation of a product cannot be detected.

EXERCISES

1. 8·0 g of oxygen, sealed in a closed vessel, is heated electrically, and a rise in temperature of 1·33°C is noted. If the electrical heater supplied a total of 7 J of energy, calculate C_v, C_p and γ for oxygen. Take R as 8·3 J mole^{-1} K^{-1}.
2. Calculate the rise in temperature of 22 g of carbon dioxide which receives 100 J of heat at (a) constant volume, (b) constant pressure, given that C_p and γ for the gas are 37·2 J mole^{-1} K^{-1} and 1·3 respectively.
3. What is the heat produced by burning 100 litres of carbon monoxide (measured at s.t.p.), given the equation

   $$CO + \tfrac{1}{2}O_2 = CO_2 \quad \Delta H = -285 \text{ kJ}$$

   and the fact that one mole of gas occupies 22·4 litres at s.t.p.
4. Using the data below, show that more heat is liberated when carbon burns in dinitrogen oxide, N_2O, than in pure oxygen, and explain why this should be so.

$$C + O_2 = CO_2 \quad \Delta H = -394 \text{ kJ}$$
$$N_2 + \tfrac{1}{2}O_2 = N_2O \quad \Delta H = 73 \text{ kJ}$$

5. Deduce the heat of formation of carbon disulphide from the following data:

$$C + O_2 = CO_2 \quad \Delta H = -403 \text{ kJ}$$
$$S + O_2 = SO_2 \quad \Delta H = -298 \text{ kJ}$$
$$CS_2 + 3O_2 = 2SO_2 + CO_2 \quad \Delta H = -1113 \text{ kJ}$$

6. Calculate the heats of formation of ethane and ethene from the following values:

$$C + O_2 = CO_2 \quad \Delta H = -403 \text{ kJ}$$
$$H_2 + \tfrac{1}{2}O_2 = H_2O \quad \Delta H = -285 \text{ kJ}$$
$$C_2H_4 + 3O_2 = 2CO_2 + 2H_2O \quad \Delta H = -1395 \text{ kJ}$$
$$C_2H_6 + 3\tfrac{1}{2}O_2 = 2CO_2 + 3H_2O \quad \Delta H = -1554 \text{ kJ}$$

7. If 2 moles of hydrogen peroxide are decomposed catalytically at s.t.p., to produce 1 mole of gaseous oxygen, calculate the work of expansion.

8. Using the heats of formation calculated in answer to Question 6, show that the heat of hydrogenation of ethene is $\Delta H = -126$ kJ.

9. If the heats of formation of aluminium (III) oxide and chromium (III) oxide are 1596 kJ and 1134 kJ (both exothermic) respectively, calculate ΔH for the reaction

$$2Al + Cr_2O_3 = 2Cr + Al_2O_3$$

10. For the reaction

$$CO + 2H_2 = CH_3OH \quad \Delta H = -90 \cdot 3 \text{ kJ}, \Delta S = -218 \text{ J K}^{-1}$$

calculate ΔG at 300 K and at 500 K, and comment on the results.

11. Find the heat of formation of propane from the following data:

$$C + O_2 = CO_2 \quad \Delta H = -399 \text{ kJ}$$
$$2H_2 + O_2 = 2H_2O \quad \Delta H = -571 \text{ kJ}$$
$$C_3H_8 + 5O_2 = 3CO_2 + 4H_2O \quad \Delta H = -2200 \text{ kJ}$$

12. Using the following data, calculate the lattice energy of lithium fluoride.

$$Li_s + \tfrac{1}{2}F_{2,g} = Li^+F^-_s \quad \Delta H_f = -614 \text{ kJ}$$
$$Li_s = Li^+_g + e^- \quad \Delta H = 683 \text{ kJ}$$
$$\tfrac{1}{2}F_{2,g} + e^- = F^-_g \quad \Delta H = -333 \text{ kJ}$$

If the first ionization energy of lithium is 520 kJ mole^{-1} and the bond dissociation energy of fluorine is 162 kJ mole^{-1} calculate also (a) the heat of atomization of lithium and (b) the electron affinity of fluorine.

13. Evaluate, using the Born equation, the lattice energy of potassium bromide. The value for the Madelung constant is 1·747 and the ionic radii are K^+ 1·33 Å and Br^- 1·96 Å.
14. Calculate the heat of formation of the hypothetical ionic compound Ar^+Cl^-, assuming that the lattice energy is close to that of K^+Cl^- (say 700 kJ mole^{-1}). The electron affinity of chlorine is 366 kJ mole^{-1}, the bond dissociation energy of chlorine is 244 kJ mole^{-1} and the first ionization energy of argon is 1517 kJ mole^{-1}. Comment on the result.
15. Evaluate the electron affinity of iodine if the lattice energy and heat of formation of sodium iodide are 690 kJ mole^{-1} and -290 kJ mole^{-1} respectively. The following data are also required:

$$Na_s = Na^+_g + e^- \quad \Delta H = 600 \text{ kJ mole}^{-1}$$
$$\tfrac{1}{2}I_{2,s} = I_g \quad \Delta H = 105 \text{ kJ mole}^{-1}$$

16. If the heat of atomization of carbon is 716 kJ mole^{-1} and the bond dissociation energy per mole of chlorine gas is 243 kJ, calculate the average C—Cl bond energy in tetrachloromethane. The heat of formation of gaseous tetrachloromethane is

$$\Delta H_f = -113 \text{ kJ mole}^{-1}.$$

17. If the heat of atomization of C_2Cl_6 is 2306 kJ mole^{-1}, use the value for E_{C-Cl} obtained in Exercise 16 along with these data to estimate the C—C bond energy. (Assume the bond energies are additive.)
18. The heat of formation of ethyne is $\Delta H_f = +209$ kJ mole^{-1}. Compare this value with that calculated from bond energies, given:

$E_{C-H} = 416$ kJ mole^{-1}, $E_{C\equiv C} = 815$ kJ mole^{-1},
$D_{H-H} = 432$ kJ mole^{-1}

and the heat of atomization of carbon is 716 kJ mole^{-1}. Is the structure H—C≡C—H in accordance with the results?
19. Using the bond energy data given at the rear of the book (p. 244) calculate the pure covalent bond energies of E_{C-Cl} and E_{C-Br} and compare these with the values of 327 kJ mole^{-1} and 272 kJ mole^{-1} found experimentally.
20. Benzene boils at 80°C and the latent heat of vaporization is 30·8 kJ mole^{-1}. Deduce the change in entropy corresponding to the vaporization of one mole of benzene at 80°C.
21. The following data relate to 25°C and one atmosphere pressure.

$Mg + \tfrac{1}{2}O_2 = MgO \quad \Delta H = -600$ kJ, $\Delta S = -109$ J K^{-1}
$C + \tfrac{1}{2}O_2 = CO \quad \Delta H = -120$ kJ, $\Delta S = +90$ J K^{-1}

Calculate ΔG for the reaction
$$MgO + C = Mg + CO$$
at (a) 25°C and (b) 3000 K, and say whether reduction of magnesium oxide by carbon is feasible at the latter temperature. (Assume ΔH and ΔS remain constant over this temperature range and ignore any change due to a change of state of magnesium metal.)

5. Chemical Reactions and Equilibrium

A chemical reaction may be expressed by the general equation

Reactants → Products

However, the stoichiometric equation (i.e. an equation such as

$$PCl_5 = PCl_3 + Cl_2,$$

which identifies the species and the number of moles of each species taking part in the reaction) does not give any information about:

(a) whether the reaction could take place or not,
(b) the quantity of product obtained from a given quantity of reactant,
(c) the velocity of the reaction,
(d) the stages (or mechanism) by which the reaction proceeds.

The answer to item (a) regarding the possibility of a reaction taking place is given in terms of free energies, as outlined in the previous chapter. The question of the yield of a reaction is discussed under the heading of chemical equilibrium in this chapter, Items (c) and (d) concerning the velocity and mechanism of a reaction are also discussed later in this chapter under the heading of the kinetics of chemical change.

Chemical Equilibrium

When heat is given to water at 100°C (at 1 atm), some steam is formed and the temperature remains constant. If the conditions are held at this point, we have a system in which the water and steam are in *equilibrium*. If more heat is added, more steam is produced while, on the removal of heat, some steam condenses to water. The system is said to be *reversible*, that is, by reversing the conditions we can reverse the process taking place in the system. Similarly, there are many chemical reactions which are reversible. The esterification of ethanol is a typical example:

$$C_2H_5OH + CH_3COOH \rightleftharpoons CH_3COOC_2H_5 + H_2O$$

Since the reaction can take place in both directions, a point must come when the rate of formation of ester is equal to the rate of its decomposition. Such reversible (or balanced) reactions ultimately produce an equilibrium state, more accurately described as a *dynamic equilibrium* since it represents a state of balance between the forward and backward reactions. It is obvious that changes in many factors, such as temperature, pressure, concentration, presence of catalysts and so on can alter this balanced state; hence we say that the equilibrium is readily displaced. Often, we regard a reaction as going to completion, meaning that the amount of the reactants remaining unused at the end of the reaction falls below the level of detection of our analytical techniques. These reactions can still be considered (at least theoretically) as reversible reactions in which the equilibrium has been displaced to the extreme right.

Rate of reaction The rate at which a reaction takes place can be expressed as the decrease in the quantity of reactants, or the increase in the quantity of products per unit time.

Expressed mathematically, if

$$\text{Reactants} \rightarrow \text{Products}$$

number of moles $\quad n_r \quad n_p$

$$\text{rate of reaction} = \frac{-dn_r}{dt} = \frac{dn_p}{dt}$$

The law of mass action (Guldberg and Waage 1864) *The rate at which a chemical reaction takes place, at a given temperature, is proportional to the product of the active masses of the reactants.*

The active mass of a substance dissolved in dilute solution is equal to its molar concentration. This is designated by writing [A], where

$$[A] = \frac{\text{number of moles of A}}{\text{volume of solution in litres}}$$

The active mass of a gas present in a mixture is equal to its partial pressure, while the active mass of a pure solid is taken as a constant value, irrespective of the amount of the solid present.

For a reaction

$$A + B \rightarrow \text{Products}$$

the rate of reaction is proportional to $[A] \times [B]$ in solution or proportional to $p_A \times p_B$ in a gas mixture. Hence

$$\text{rate of reaction} = k[A][B] \quad \text{in solution}$$

where k, the constant of proportionality, is termed the *velocity constant* for the reaction. (Similarly,

$$\text{rate of reaction} = k' \times p_A \times p_B \quad \text{for gaseous reactions.})$$

For the general reaction

$$x\text{A} + y\text{B} \to \text{products}$$

we have

$$\text{rate of reaction} = k[\text{A}]^x[\text{B}]^y$$

For a reversible reaction

$$\text{A} + \text{B} \rightleftharpoons \text{C} + \text{D}$$

we have

$$\text{rate of forward reaction} = k[\text{A}][\text{B}]$$
$$\text{rate of reverse reaction} = k'[\text{C}][\text{D}]$$

At equilibrium, the rates of the two reactions balance, so that

$$k[\text{A}][\text{B}] = k'[\text{C}][\text{D}]$$

or

$$\frac{k}{k'} = K_c = \frac{[\text{C}][\text{D}]}{[\text{A}][\text{B}]}$$

K_c is called the equilibrium constant for the reaction (expressed in terms of concentrations). Although k'/k is also a constant, K_c is written so as to equal the ratio of the velocity constants of the forward and reverse reactions.

For a gaseous reaction $\text{A} + \text{B} \rightleftharpoons \text{C} + \text{D}$,

$$K_p = \frac{p_C \times p_D}{p_A \times p_B}$$

and K_p is the equilibrium constant expressed in terms of partial pressures.

Reversible reactions which take place in one phase only (i.e. liquid or gaseous) lead to the formation of *homogeneous equilibria*. If more than one phase (e.g. a solid and a gas) is present, a *heterogeneous equilibrium* is set up.

Homogeneous equilibria In a reaction of the type

$$\text{A} + \text{B} \rightleftharpoons \text{C} + \text{D}$$

if a moles of A and b moles of B were initially present in a volume of V litres; and at equilibrium, x moles of A and B have reacted to form x moles of C and x moles of D, then the equilibrium concentrations of the various species are:

$$[A] = \frac{a-x}{V} \quad [B] = \frac{b-x}{V} \quad [C] = \frac{x}{V} \quad [D] = \frac{x}{V}$$

and

$$K_c = \frac{\frac{a-x}{V} \times \frac{b-x}{V}}{\frac{x}{V} \times \frac{x}{V}} = \frac{(a-x) \times (b-x)}{x^2}$$

Example 50 7·2 g of water and 2 moles of ethyl acetate were refluxed until an equilibrium was established. If the equilibrium mixture contains 0·38 mole each of acetic acid and ethanol, calculate the equilibrium constant for the reaction

$$CH_3COOC_2H_5 + H_2O \rightleftharpoons CH_3COOH + C_2H_5OH$$

Solution Assuming the volume of the system is V litres, then the concentrations of the constituents at equilibrium are:

$$[\text{ethyl acetate}] = \frac{2 - 0·38}{V} = \frac{1·62}{V} \text{ mole l}^{-1}$$

$$[\text{water}] = \frac{7·2/18 - 0·38}{V} = \frac{0·02}{V} \text{ mole l}^{-1}$$

$$[\text{acetic acid}] = \frac{0·38}{V} \text{ mole l}^{-1}$$

$$[\text{ethanol}] = \frac{0·38}{V} \text{ mole l}^{-1}$$

$$K_c = \frac{[CH_3COOH] \times [C_2H_5OH]}{[CH_3COOC_2H_5] \times [H_2O]} = \frac{\left(\frac{0·38}{V}\right) \times \left(\frac{0·38}{V}\right)}{\left(\frac{1·62}{V}\right) \times \left(\frac{0·02}{V}\right)} = 4·46$$

Note that in this case K_c has no units.

The hydrogen–iodine reaction If the initial quantities of hydrogen and iodine are a and b moles respectively, and, at equilibrium, x moles of each have reacted to form hydrogen iodide, then the concentrations of each constituent at equilibrium are:

$$H_2 + I_2 \rightleftharpoons 2HI$$

$$\frac{a-x}{V} \quad \frac{b-x}{V} \quad \frac{2x}{V}$$

if the total volume is V litres. Thus

$$K_c = \frac{[HI]^2}{[H_2] \times [I_2]} = \frac{4x^2}{(a-x) \times (b-x)}$$

Bodenstein, in 1897, used this reaction to demonstrate the validity of the principles of equilibrium. Known quantities of hydrogen and iodine were sealed in hard glass tubes and kept for several hours at 444°C. The tubes were then cooled rapidly so that the equilibrium did not have time to alter to the new conditions (reactions take place much more slowly at lower temperatures so that any readjustment of the equilibrium takes a longer period—this is known as *freezing the equilibrium*). The contents of the several tubes were analysed and a mean value for the equilibrium constant was calculated. Bodenstein then took widely varying quantities of hydrogen and iodine and repeated the experiment. In each case, the quantity of hydrogen iodide found to be present at equilibrium was compared with the quantity predicted by calculation from the value of the equilibrium constant previously determined. The two sets of values were found to correspond closely. Finally, the whole experiment was repeated using hydrogen iodide as the starting material, and the same conclusions were reached.

Example 51 At 400°C, 2·4 moles of hydrogen, 5·2 moles of iodine and 26·5 moles of hydrogen iodide were found to be present in an equilibrium mixture. Calculate the value of the equilibrium constant at this temperature.

Solution If a moles of hydrogen and b moles of iodine were present initially, the equilibrium quantities are:

$$\begin{array}{ccc} H_2 & + \quad I_2 & \rightleftharpoons \quad 2HI \\ a-x & b-x & 2x \end{array}$$

in a volume of V litres. But $a-x = 2·4$, $b-x = 5·2$, while $2x$, the total number of moles of hydrogen iodide, is 26·5. Thus

$$K_c = \frac{\left(\frac{2x}{V}\right)^2}{\frac{a-x}{V} \times \frac{b-x}{V}} = \frac{(26·5)^2}{2·4 \times 5·2} = 56·3$$

The thermal decomposition of phosphorus pentachloride In both the esterification equilibrium and the hydrogen–iodine reaction the volume term cancels in the expression for the equilibrium constant.

This happens only when equal numbers of molecules of reactants and products are involved. The thermal decomposition of phosphorus pentachloride is an example in which the volume term is important.

At equilibrium, the number of moles of each constituent are given by:

$$PCl_5 \rightleftharpoons PCl_3 + Cl_2$$
$$a-x \quad x \quad x$$

in a volume of V litres, a being the amount of phosphorus pentachloride initially present. Thus

$$K_c = \frac{[PCl_3] \times [Cl_2]}{[PCl_5]} = \frac{\left(\frac{x}{V}\right) \times \left(\frac{x}{V}\right)}{\frac{a-x}{V}} = \frac{x^2}{V(a-x)} \text{ mole l}^{-1}$$

Example 52 When one mole of phosphorus pentachloride vapour is heated to 250°C in a 50-litre vessel, the equilibrium mixture is found to contain 0·5 mole of chlorine. Calculate the value of the equilibrium constant, and deduce the effect of doubling the pressure on the equilibrium mixture.

Solution The number of moles of each constituent at equilibrium are given by:

$$PCl_5 \rightleftharpoons PCl_3 + Cl_2$$
$$1-x \quad x \quad x$$

x has the value 0·5, which makes $1-x$ equal to 0·5 also. Substituting

$$K_c = \frac{x^2}{V(1-x)} = \frac{0\cdot5 \times 0\cdot5}{50 \times 0\cdot5} = 0\cdot01 \text{ mole l}^{-1}$$

If the pressure is doubled, the volume is halved, assuming that Boyle's law is obeyed. Hence,

$$0\cdot01 = \frac{x^2}{25(1-x)}$$

Multiplying out and re-arranging,

$$x^2 + 0\cdot25x - 0\cdot25 = 0$$

Solving,*

$$x = \frac{-0\cdot25 \pm \sqrt{(0\cdot0625 + 1)}}{2} = 0\cdot39 \text{ or } -0\cdot64$$

* The solution to a quadratic equation $ax^2 + bx + c = 0$ is given by

$$x = \frac{-b \pm \sqrt{(b^2 - 4ac)}}{2a}$$

A negative value of x is not a solution to this problem, so $x = 0.39$. Therefore, the equilibrium mixture contains 0·39 moles of chlorine, 0·39 moles of phosphorus trichloride, and 0·61 moles of phosphorus pentachloride. Note that the increase in pressure has resulted in a greater proportion of phosphorus pentachloride being present at equilibrium; this is in accordance with Le Chatelier's principle (p. 153).

Equilibrium in Gaseous Systems

In gaseous systems, the active mass of each constituent may be expressed in terms of partial pressures. For example, for the equilibrium

$$N_2 + 3H_2 \rightleftharpoons 2NH_3$$

$$K_p = \frac{(p_{NH_3})^2}{(p_{N_2}) \times (p_{H_2})^3}$$

Assuming that one mole of nitrogen and three moles of hydrogen were present initially, the number of moles of each species at equilibrium can be represented as follows:

$$\begin{array}{cccc} N_2 & + & 3H_2 & \rightleftharpoons & 2NH_3 \\ 1-x & & 3-3x & & 2x \end{array}$$

The total number of moles present at equilibrium is

$$(1-x) + (3-3x) + 2x = 4 - 2x$$

If the total pressure is P, the partial pressures at equilibrium are:

$$p_{N_2} = \frac{1-x}{4-2x} \times P \qquad p_{H_2} = \frac{3-3x}{4-2x} \times P \qquad p_{NH_3} = \frac{2x}{4-2x} \times P$$

$$K_p = \frac{(p_{NH_3})^2}{(p_{N_2}) \times (p_{H_2})^3} = \frac{\dfrac{4x^2 P^2}{(4-2x)^2}}{\left(\dfrac{1-x}{4-2x}\right) P \times \left(\dfrac{3-3x}{4-2x}\right)^3 P^3}$$

$$= \frac{4x^2(4-2x)^2}{27\, P^2 (1-x)^4}$$

Example 53 The equilibrium mixture formed by heating one mole of nitrogen and three moles of hydrogen at 50 atmospheres was found to contain 0·8 mole of ammonia. Deduce the value of K_p under these conditions.

Solution The number of moles of each constituent present at equilibrium is given by:

$$N_2 + 3H_2 \rightleftharpoons 2NH_3$$
$$1-x \quad 3-3x \quad 2x$$

Since the number of moles of ammonia present at equilibrium is 0·8,

$$2x = 0·8 \quad \text{or} \quad x = 0·4$$

Substituting for x and P in the above equation:

$$K_p = \frac{4x^2(4-2x)^2}{27\,P^2(1-x)^4} = \frac{4 \times 0·4^2 \times 3·2^2}{27 \times 50^2 \times 0·6^4}$$

$$= 9·4 \times 10^{-4} \text{ atm}^2$$

Displacement of Equilibrium: Factors Affecting Chemical Equilibrium

The establishment of a chemical equilibrium is the result of a balance between the forward and reverse reactions of a reversible reaction. Therefore the equilibrium position is altered by factors which affect the rates of the forward and reverse reactions unequally, and their effect was summarized by H. le Chatelier (1885).

Le Chatelier's principle of mobile equilibrium *If one of the external factors which affect a system in equilibrium is altered, then the system tends to react in such a way as to minimize any changes resulting from the alteration of the external factors.*

Displacement caused by concentration changes If we consider the general reaction,

$$A + B \rightleftharpoons C + D$$

for which

$$K_c = \frac{[C][D]}{[A][B]}$$

we can see that, provided the total volume and other factors such as temperature are kept constant, an increase in the concentration of (say) A will produce an increase in the concentrations of C and D, since the constant value for K_c must be preserved.

Example 54 The equilibrium constant for the formation of ethyl acetate is 4·0 at 70°C. What concentration of ester would be produced at equilibrium starting from one mole of acetic acid and (a) 2 moles of alcohol, (b) 10 moles of alcohol?

Solution $\quad C_2H_5OH + CH_3COOH \rightleftharpoons CH_3COOC_2H_5 + H_2O$

No. of moles
present initially $\quad\quad a \quad\quad\quad 1 \quad\quad\quad$ zero $\quad\quad$ zero

At equilibrium $\quad a-x \quad\quad 1-x \quad\quad\quad x \quad\quad\quad x$

(a) When $a = 2$,

$$K = 4 = \frac{x^2}{(1-x)(2-x)}$$

$$3x^2 - 12x + 8 = 0$$

$$x = 0\cdot 84 \text{ mole}$$

(b) When $a = 10$,

$$K = 4 = \frac{x^2}{(1-x)(10-x)}$$

$$3x^2 - 44x + 40 = 0$$

$$x = 0\cdot 97 \text{ mole}$$

Thus, a greater yield of ester follows an increase in the initial concentration of alcohol (or acid), and the equilibrium is displaced to the right as shown in Figure 53. If one of the products is removed

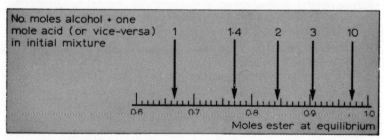

Fig. 53. Influence of concentration on the position of equilibrium

from the reaction medium as it is formed (either by distillation, precipitation or solvent extraction, etc.) then the equilibrium is displaced so far that the reaction may go to completion. A system in which a product is removed in this way is referred to as an *open system*, whereas we are confining our discussions to *closed systems* in which all the products and reactants are retained within the reaction medium.

Displacement caused by changes in pressure Pressure affects only those equilibria which show a change in volume between the reactants and the products. The effects of pressure changes on reactions in which gases are involved are particularly important, for the following reasons:

(a) In any gaseous reaction, an increase in pressure brings the reacting molecules closer together, and the reaction speeds up. This factor affects both forward and reverse reactions.
(b) The composition of the equilibrium mixture (but not the equilibrium constant) is altered in those cases where there is a volume change (see Example 52 and p. 152).

For instance, in the reaction

$$2SO_2 + O_2 \rightleftharpoons 2SO_3$$

the gaseous reactants occupy three volumes to the product's two. To counterbalance the effect of an increase in pressure, the system will take the form which occupies a smaller volume as the pressure is applied—that is, more sulphur trioxide will be formed.

Displacement caused by changes in temperature In all reactions, an increase in temperature causes an increase in the speed at which a reaction takes place. The rate of reaction approximately doubles or trebles for every 10°C rise in temperature. This means that, at higher temperatures, the equilibrium position is attained more rapidly.

Temperature also has an effect on the *value* of the equilibrium constant. It may be shown that

$$\log K \propto -\Delta H / T$$

where ΔH is the heat of reaction. Hence, for an exothermic reaction in which ΔH is negative, $\log K$ (and therefore K) will become progressively smaller as the temperature increases. Since

$$K = \frac{[\text{products}]}{[\text{reactants}]}$$

the product concentration will fall with increasing temperature.

For endothermic reactions, ΔH is positive, and at any temperature, $\log K$ is negative. Thus,

$$\log K \propto \frac{-\Delta H}{T}$$

or,

$$\log \frac{1}{K} \propto \frac{\Delta H}{T}$$

Hence, as the temperature increases, K increases and the product concentration will rise.

The same conclusions may be arrived at directly from le Chatelier's theorem. If the forward reaction in the equilibrium is exothermic, heat is lost in forming the products. Since it is easier to dissipate heat at lower temperatures, it is expected that the quantity of product formed will be greater at lower temperatures. The converse is true when the forward reaction is endothermic.

Effect of catalysts A catalyst speeds up the rate of a chemical reaction. In equilibrium reactions, both forward and reverse reactions are equally affected, so that the position of equilibrium is not changed, although the equilibrium state is arrived at more rapidly.

Equilibrium Reactions of Industrial Importance

The Haber process The production of ammonia from nitrogen and hydrogen is exothermic, and is accompanied by a decrease in volume:

$$3H_2 + N_2 \rightleftharpoons 2NH_3 \quad \Delta H = -92 \text{ kJ}$$

It follows that a better yield of ammonia may be obtained by working at high pressures and low temperatures. However, the influence of temperature is more complex than is at first apparent, since working at low temperatures means that the equilibrium state is established slowly. A compromise, or *optimum* temperature (i.e. one at which it is possible to obtain a reasonable yield of ammonia in a reasonable time) is therefore chosen. In practice, ammonia is produced by this reaction at 400 atmospheres and at 500°C in the presence of a catalyst.

The contact process When sulphur dioxide reacts with oxygen to form sulphur trioxide, a decrease in volume is noted, and heat is evolved:

$$2SO_2 + O_2 \rightleftharpoons 2SO_3 \quad \Delta H = -188 \text{ kJ}$$

As with the ammonia reaction, high pressures and lower temperatures increase the yield of sulphur trioxide. To give a reasonable rate of reaction, an optimum temperature of 450°C and a catalyst are employed.

The Ostwald process This is the modern method for the manufacture of nitric acid by the catalytic oxidation of ammonia:

$$4NH_3 + 5O_2 \rightleftharpoons 4NO + 6H_2O \quad \Delta H = -900 \text{ kJ}$$

In this reaction there is a small increase in volume, and it would be expected that low pressures would be used. However, as mentioned on p. 155, an increase in pressure has the effect of speeding up a reaction by bringing the molecules closer together, and, of course, smaller sized plant can be employed. In practice, this process operates at pressures up to 9 atmospheres and at temperatures between 600° and 1000°C. A platinum–rhodium catalyst is used, and an excess of air or oxygen is admitted, which also helps to give a greater amount of nitrogen monoxide in the final mixture.

Heterogeneous Equilibria

When the reactants or products are present in more than one phase, the equilibrium is said to be *heterogeneous*, and difficulties arise in deciding what represents the activity of a substance in various phases, especially in the solid phase. The law of mass action is usually applied to heterogeneous equilibria by assuming the active mass of a solid to be a constant at a given temperature (and independent of the amount of solid present). For example:

$$CaCO_{3,s} \rightleftharpoons CaO_s + CO_{2,g}$$

$$K = \frac{[CaO] \times [CO_2]}{[CaCO_3]}$$

which, if the active masses of calcium carbonate and calcium oxide are taken as constants, reduces to

$$K \propto [CO_2] \quad \text{or} \quad K_p \propto p_{CO_2}$$

since the partial pressure of the carbon dioxide can be used to express its concentration.

Similarly, in the equilibrium

$$3Fe_s + 4H_2O_g \rightleftharpoons Fe_3O_{4,s} + 4H_{2,g}$$

$$K = \frac{[H_2]^4 \times [Fe_3O_4]}{[Fe]^3 \times [H_2O]^4}$$

which reduces to

$$K_p \propto \frac{p_{H_2}}{p_{H_2O}}$$

That is, the ratio of the partial pressures of hydrogen and steam is constant for all proportions of iron and iron oxide at a given temperature.

Thermal Dissociation

Many substances decompose on heating, and, provided the products of decomposition are not removed, will recombine again on cooling. This is known as *thermal dissociation*. For example,

$$AB \underset{\text{cool}}{\overset{\text{heat}}{\rightleftharpoons}} A + B$$

For any given temperature, an equilibrium is set up. Typical examples are:

$$NH_4Cl \rightleftharpoons NH_3 + HCl$$
$$PCl_5 \rightleftharpoons PCl_3 + Cl_2$$
$$N_2O_4 \rightleftharpoons 2NO_2$$

The course of these reactions can be followed by measurements of relative (vapour) density since, on dissociation, a greater total volume is occupied by the system, resulting in a fall in density.

The *degree of dissociation* (α) is the fraction of a mole of the starting material which has dissociated (α may also be expressed as a percentage).

Example 55 If the relative density of dinitrogen tetroxide at 62°C is 30·0, calculate the degree of dissociation.

Solution The number of moles of each constituent at equilibrium is given by:

$$\begin{array}{cc} N_2O_4 \rightleftharpoons & 2NO_2 \\ 1-\alpha & 2\alpha \end{array}$$

The total number of moles present at equilibrium is

$$(1-\alpha) + 2\alpha = 1 + \alpha$$

The volume occupied by the system is proportional to $1+\alpha$, and so the observed density of the system is proportional to $1/(1+\alpha)$. The density if no dissociation has taken place is proportional to $1/1$. Thus

$$\frac{\text{density with no dissociation}}{\text{observed density}} = \frac{1+\alpha}{1} = \frac{\frac{2 \times 14 + 4 \times 16}{2}}{30}$$

Hence
$$1+\alpha = 1\cdot 53$$
$$\alpha = 0\cdot 53 \text{ or } 53\%$$

Immiscible Solvents

A type of equilibrium is set up when a substance is added to a pair of immiscible liquids, the substance being soluble in both solvents (Figure 54). The equilibrium is described by the distribution, or partition, law.

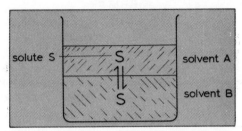

Fig. 54. Partition of a solute between two immiscible solvents

The distribution law *At a given temperature, the solute distributes itself in such a way that the ratio of the concentrations of the solute in each layer is a constant, provided the solute is present in the same molecular species in each phase.*

Stated mathematically, the partition law takes the form:

$$\frac{\text{concentration of solute in solvent A}}{\text{concentration of solute in solvent B}} = K$$

where K is the *partition* or *distribution coefficient*.

The following values resulting from the distribution of succinic acid between ether and water illustrate the constancy of the ratio of the concentrations in each layer.

Concentration of acid in ether (C_1)	Concentration of acid in water (C_2)	$K = \dfrac{C_1}{C_2}$
0·0046	0·024	0·191
0·013	0·069	0·188
0·022	0·119	0·185
0·031	0·164	0·189

Example 56 A solid X is added to a mixture of benzene and water. After shaking well, 10 cm^3 of the benzene layer was found to contain

0·13 g of X, while 100 cm³ of the water layer contained 0·22 g of X. Calculate the value of the partition coefficient.

Solution

$$\text{Concentration of X in benzene} = \frac{0·13}{10} = 0·013 \text{ g cm}^{-3}$$

$$\text{Concentration of X in water} = \frac{0·22}{100} = 0·0022 \text{ g cm}^{-3}$$

and

$$K = \frac{0·013}{0·0022} = 5·9$$

Association of the solute In the cases where the solute undergoes association in one of the solvents, the partition law is modified. Suppose the solute associates into double molecules in solvent A (Figure 55). Assuming that few single molecules of S are present in

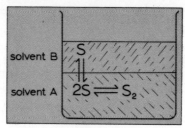

Fig. 55

solvent A, then the concentration of S in that solvent refers to double molecules (S_2). For the equilibrium established in A,

$$\frac{[S_2]}{[S]^2} = \text{a constant (say } K')$$

Thus the concentration of single molecules is given by:

$$[S] \propto \sqrt{(\text{concentration of S in A})}$$

But the partition coefficient refers to the equilibrium between single molecules of S in A and B, i.e.,

$$K = \frac{\text{concentration of single molecules of S in A}}{\text{concentration of single molecules of S in B}}$$

$$= \frac{\sqrt{(\text{concentration of the solute in A})}}{\text{concentration of S in B}}$$

Therefore, if association into double molecules takes place in one of the solvents, the square root of the concentration of the solute in that solvent must be used in calculating the value of the partition coefficient.

Example 57 When benzoic acid was shaken with mixtures of benzene and water, at a constant temperature, the following results were obtained.

Equilibrium concentration of benzoic acid in benzene (C_1)	0·24	0·55	0·93
Equilibrium concentration of benzoic acid in water (C_2)	0·015	0·022	0·029

Comment on these results.

Solution Calculating the ratio C_1/C_2 for each case:

$$C_1/C_2 = \frac{0\cdot24}{0\cdot015} = 16$$

$$= \frac{0\cdot55}{0\cdot022} = 25$$

$$= \frac{0\cdot93}{0\cdot029} = 32$$

and a constant value does not result. But calculating $\sqrt{C_1}/C_2$ gives:

$$\frac{\sqrt{(0\cdot24)}}{0\cdot015} = 32\cdot6$$

$$\frac{\sqrt{(0\cdot55)}}{0\cdot022} = 33\cdot7$$

$$\frac{\sqrt{(0\cdot93)}}{0\cdot029} = 33\cdot2$$

The agreement between these figures suggests that benzoic acid is associated into double molecules in the benzene layer.

Solvent Extraction

This is the process often carried out to extract organic materials from aqueous solution. Ether is commonly used as the extracting solvent, and Example 58 shows that it is more efficient to use a given volume of organic solvent in small lots, rather than in one whole.

Example 58 100 cm³ of an aqueous solution contains 10 g of an organic nitrophenol. Calculate the weight of nitrophenol extracted by 100 cm³ of ether, used (a) in one extraction, (b) in four separate extractions, using 25 cm³ of ether for each extraction. The partition coefficient of the nitrophenol between ether and water is 3:1 at room temperature.

Solution Suppose x g of the nitrophenol are extracted into the ether layer at equilibrium.

(a) Using all the ether in one portion,

$$K = 3 = \frac{x/100}{(10-x)/100}$$

from which, $x = 7 \cdot 5$ g, or 75% of the nitrophenol has been extracted.

(b) Using four separate 25 cm³ portions of ether,

$$K = 3 = \frac{x_1/25}{(10-x_1)/100}$$

from which $x_1 = 4 \cdot 28$ g. This leaves $10 - 4 \cdot 28 = 5 \cdot 72$ g nitrophenol in the water layer.

In the second extraction,

$$K = 3 = \frac{x_2/25}{(5 \cdot 72 - x_2)/100}$$

from which $x_2 = 2 \cdot 45$ g. In a similar way, it can be shown that $x_3 = 1 \cdot 40$ g and $x_4 = 0 \cdot 80$ g. The total quantity of the nitrophenol extracted in this way is $x_1 + x_2 + x_3 + x_4 = 8 \cdot 93$ g or 89·3%.

Kinetics of Chemical Change

A chemical reaction takes place:

(a) when molecules collide,
(b) if sufficient energy is liberated on collision,
(c) if the orientation of the colliding molecules (so called collision geometry) is such that the bonds most likely to break are in favourable positions for breaking and reforming at the moment of impact.

Distribution of energy For a monatomic gas, the distribution of kinetic energy among the molecules was stated on p. 33 to be expressed by a Maxwell distribution curve (Figure 56). The number n of molecules out of a total of N molecules having a kinetic energy greater than a given value E at any temperature T K is given by

$$n/N = e^{-E/RT}$$

Hence, as the temperature rises, the kinetic energy increases and the proportion of molecules having energy E rises. On collision, bonds break and new ones are formed. In some cases, it is possible to deduce that the initial step in the reaction is bond fission:

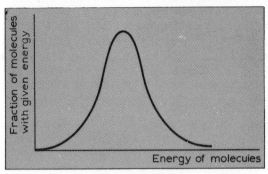

In other reactions, it appears that a new bond is formed first:

$$X-Y+Z \longrightarrow X \ldots Y-Z \longrightarrow X+Y-Z$$

Fig. 56. Maxwell distribution curve

This fine detail corresponding to the sequence of stages in a reaction is referred to as the *mechanism* of the reaction, and in both cases, the intermediate state is known as the *transition state* or

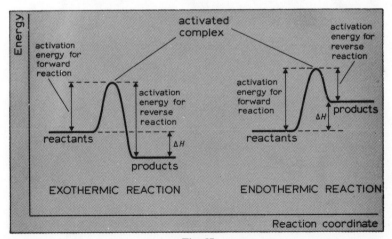

Fig. 57

activated complex. In both cases energy, known as activation energy, is required to produce the transition state. The energy diagrams for both exothermic and endothermic reactions are shown in Figure 57.

At higher temperatures, collisions will take place more frequently, and the proportion of molecules having an energy equal to or greater than the activation energy will increase; therefore, the rate of reaction speeds up as the temperature rises.

Collision geometry Although energy in excess of the activation energy may be released on impact, the collision of two molecules may not always be fruitful (i.e. lead to the formation of products). This is due to the fact that, on collision, the molecules are not positioned in such a manner that the energy released is transferred directly to the bonds which are about to undergo fission. Such a poorly orientated system is said to have an unfavourable collision geometry. This is illustrated in Figure 58, in which favourable and unfavourable cases of collision geometries are compared.

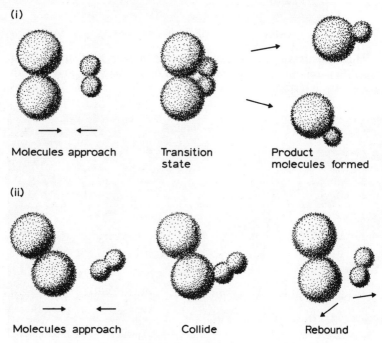

Fig. 58. Collision geometry (i) favourable (ii) unfavourable

Some reactions, especially those catalysed by enzymes, require the reacting molecules to approach in a well-defined orientation. It is clear, then, that in addition to the release of energy being at least equal to the activation energy, the orientation of the molecules on collision also plays an important part in the process.

The Arrhenius equation The proportion of molecules in a system having energy greater than a value E is given by Maxwell's equation:

$$n/N = e^{-E/RT}$$

S. Arrhenius suggested that reaction takes place only between those molecules capable of releasing an amount of energy (E) equal to the activation energy on impact. Hence

$$\text{rate of reaction} \propto k \quad \text{(the velocity constant)}$$
$$\propto e^{-E/RT}$$

or

$$k = Ae^{-E/RT}$$

where A is a constant. Taking logarithms to base e,

$$\log_e k = \text{constant} - \frac{E}{RT}$$

This is known as the Arrhenius equation in which E is the activation energy for the reaction and T is the Kelvin temperature. The important feature of the Arrhenius equation is that it describes the *change in reaction rate with temperature*. (Equations such as rate = $k[A][B]$ relate the speed of the reaction to the concentration of the reactants at a given temperature.)

Example 59 The values of k for the decomposition of diethyl ether are $2 \cdot 1 \times 10^{-4}$ l mole^{-1}s^{-1} at 480°C and 1×10^{-3} l mole^{-1}s^{-1} at 525°C. If R is 8·32 J deg^{-1} mole^{-1}, calculate the value of the activation energy for this reaction.

Solution

$$\log_e k_2 = \text{constant} - E/RT_2$$
$$\log_e k_1 = \text{constant} - E/RT_1$$

Subtracting,

$$\log_e k_2 - \log_e k_1 = \frac{E}{R}\left(\frac{1}{T_1} - \frac{1}{T_2}\right)$$

Now $\log_e k = 2 \cdot 303 \log_{10} k$ so that

$$\log k_2 - \log k_1 = \frac{E}{2 \cdot 303 R}\left(\frac{T_2 - T_1}{T_1 T_2}\right)$$

Substituting the data given,

$$\log 1 \times 10^{-3} - \log 2\cdot 1 \times 10^{-4} = \frac{E}{2\cdot 303 \times 8\cdot 32}\left(\frac{45}{753 \times 798}\right)$$

$$-3 + 4 - 0\cdot 32 = E \times 3\cdot 91 \times 10^{-6}$$

$$E = 173 \text{ kJ}$$

For many reactions (especially in solution) it is noticed that the reaction rate doubles or trebles for each 10°C rise in temperature. This agrees with the fact that the activation energy for many reactions lies in the range 50 000 to 100 000 J.

If E is 50000 J, then k at 30°C is given by

$$\exp(-50\,000/8\cdot 32 \times 303) = 2\cdot 4 \times 10^{-9} \text{ l mole}^{-1}\text{s}^{-1}$$

and at 40°C k is calculated to be $4\cdot 5 \times 10^{-9}$ l mole^{-1}s^{-1}.

Similarly, if $E = 80\,000$ J k at 10°C is $1\cdot 8 \times 10^{-15}$ l mole^{-1}s^{-1} and at 20°C it is $5\cdot 6 \times 10^{-15}$ l mole^{-1}s^{-1}.

Graph of the Arrhenius equation   If we write the Arrhenius equation

$$\log_e k = -\frac{E}{RT} + \text{constant}$$

or, using logarithms to base 10,

$$\log k = -\frac{E}{2\cdot 303 RT} + \text{constant}$$

and plot $\log k$ (on the y-axis) against $1/T$ (on the x-axis), we find the graph is a straight line of slope $-E/(2\cdot 303\ R)$. A typical case is illustrated in Example 60.

Example 60   The following data apply to the decomposition of dinitrogen pentoxide.

k l mole^{-1}s^{-1}	0·00049	0·0020	0·0300	0·3000
T°C	17	27	47	67

Estimate, graphically, the average energy of activation for this reaction.

Solution   From the given results we can construct the following table:

T K	290	300	320	340
$(1/T) \times 10^{-3}$	3·44	3·33	3·12	2·94
$\log k$	−3·31	−2·70	−1·52	−0·52

(Remember $\log 0.00049 = \bar{4}.6902 = -4+0.6902 = -3.31$)

These values are expressed graphically in Figure 59. The slope of the line taken with respect to the points shown is

$$-2.64/(0.473 \times 10^{-3}).$$

$$-\frac{2.64}{0.473 \times 10^{-3}} = -\frac{E}{2.303R}$$

from which

$$E = \frac{2.64 \times 2.303 \times 8.32 \times 10^3}{0.473} \text{ J}$$

$$= 107 \text{ kJ}$$

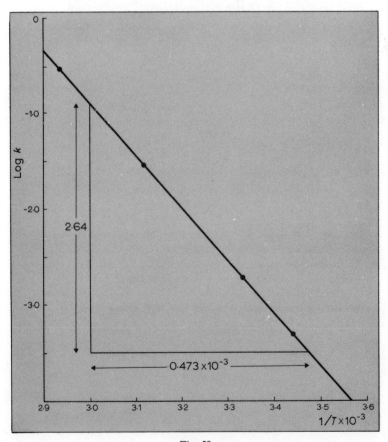

Fig. 59

The Mechanism of a Reaction

This is the elucidation of the sequence of stages by which a reaction proceeds, and a great deal of valuable information relating to reaction mechanisms is obtained from experimental chemical kinetics. If we follow the progress of a reaction, such as A → products, a graph of the concentration of A at any given instant will be similar to that shown in Figure 60.

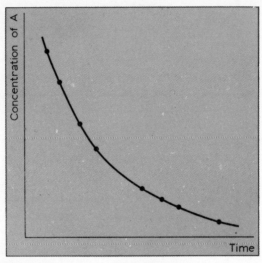

Fig. 60

The rate of reaction at any time may be expressed as the rate of disappearance of A at that moment, i.e.

$$\text{rate} = -\frac{d[A]}{dt} \quad (t = \text{time})$$

Experimental results show that for this type of reaction,

$$\text{rate of reaction} \propto [A]^a \text{ (in general)}$$

and

$$-\frac{d[A]}{dt} = k[A]^a$$

where k is the velocity constant and a is termed the *order of reaction with respect to the species A*.

If we have two reactants,

$$A + B \rightarrow \text{products}$$

the rate of reaction is in general given by
$$k[A]^a[B]^b$$
The order of the reaction with respect to A is a, and with respect to B is b, while the *overall order of reaction* is $a+b$. Thus the order of a reaction (with respect to a given species) is the power of the concentration of that species which appears in the equation determining the rate of reaction.

The importance of the order of reaction is that it allows an assessment to be made of the feasibility of an equation to represent the reaction mechanism. For example, the reaction between hydrogen gas and iodine vapour, $H_2 + I_2 \rightleftharpoons 2HI$ is found to depend on the concentration of both hydrogen and iodine to the power of one. Thus, this reaction is of first order with respect to each reactant and second order overall. This agrees with the given equation. However, in the parallel reaction
$$H_2 + Br_2 = 2HBr$$
it is found that
$$\text{the rate of reaction} \propto [Br_2]^{\frac{1}{2}}[H_2]$$
giving it an overall order of 1·5 which means that the equation cannot represent the reaction mechanism. The order of reaction need not be a whole number, and one test of a theoretical reaction mechanism is to discover whether the mathematics of the proposed mechanism gives an order of reaction which agrees with the experimental value. Quite often it is found that the commonly quoted stoichiometric equation does not indicate the order of reaction. For example, the decomposition of dinitrogen pentoxide is summarized by the equation
$$2N_2O_5 = 4NO_2 + O_2$$
Experimentally, this is found to be a first order reaction and must take place by a mechanism other than that indicated by the above equation.

The mathematics of a first order reaction The rate of a first order reaction
$$A \rightarrow \text{products}$$
is given by
$$-\frac{d[A]}{dt} = k[A]$$
Initially (at time $t = 0$) let a moles of A be present, and suppose at

time t after the start of the reaction x mole of A have decomposed, and $(a-x)$ mole remain. Assuming a volume of one litre:

At time t,

$$-\frac{d[A]}{dt} = -\frac{d(a-x)}{dt}$$

$$= \frac{dx}{dt} = k(a-x) \quad \text{(p. 168)}$$

Thus

$$dt = \frac{dx}{k(a-x)}$$

Integrating,

$$t = -\frac{1}{k}\log_e(a-x) + C$$

where C is the constant of integration.

When $t = 0$, $x = 0$ and

$$C = \frac{1}{k}\log_e a$$

Thus,

$$t = \frac{1}{k}\log_e\left(\frac{a}{a-x}\right)$$

or

$$k = \frac{1}{t}\log_e\left(\frac{a}{a-x}\right)$$

$$= \frac{2 \cdot 303}{k}\log\left(\frac{a}{a-x}\right)$$

For a given reaction, a graph of t against $\log[a/(a-x)]$ will be a straight line if the reaction is of first order. The slope of the line is $1/k$ and this is the method by which k may be estimated. Note that for a first order reaction k has dimensions of time^{-1}, while it may be shown that the dimensions of k for a second order reaction are (time^{-1})(concentration^{-1}), or units litre mole^{-1} s^{-1}.

Half-life times This is the time taken for a reaction to be half completed, i.e. for x to become $a/2$ in the above example of a first order reaction. When

$$x = \tfrac{1}{2}a, \quad t = t_{\frac{1}{2}}$$

$$t_{\frac{1}{2}} = \frac{2 \cdot 303}{k}\log\left(\frac{a}{a - \tfrac{1}{2}a}\right)$$

$$= \frac{2 \cdot 303}{k} \log 2 = \frac{0 \cdot 693}{k}$$

Note that the half-life time (or in fact the time for any other fraction of the reaction to be completed) is independent of the initial concentration of the reactant. Radioactive disintegration (p. 234) is a particularly good example of first order mathematics.

Molecularity of a reaction The molecularity of a reaction is particularly concerned with the proposed reaction mechanism. A reaction may take place in a series of stages, one of which will be the slowest. This stage is known as the *rate determining step* in the reaction, and the molecularity of a reaction is defined as the number of molecules (or atoms or radicals) which take part in the rate determining step. The molecularity of a complete step is always a whole number, and for elementary reactions (i.e. those in which a simple stoichiometric equation expresses the mechanism of the process) the molecularity and the order of the reaction are numerically equal.

Collision theory of reaction rates The dissociation of hydrogen iodide

$$2HI \rightleftharpoons H_2 + I_2$$

is a second order, bimolecular (i.e. has a molecularity of two) reaction. This means that two molecules of hydrogen iodide are involved in the collision which is the rate determining step. By measuring the variation of the velocity constant with temperature, the activation energy (E) of the reaction is found to be 184 kJ mole^{-1}. With a concentration of HI of one mole per litre at atmospheric pressure and at 283°C, the number of molecules having energy greater than the activation energy is calculated from

$$n/N = e^{-E/RT}$$

to be $5 \cdot 7 \times 10^{-18}$ per cm^3.

From kinetic theory considerations, the number of molecules per cm^3 colliding per second is $5 \cdot 9 \times 10^{31}$. Thus, the number of collisions per second between activated molecules is

$$5 \cdot 7 \times 10^{-18} \times 5 \cdot 9 \times 10^{31} = 3 \cdot 4 \times 10^{14} \text{ per cm}^3$$

Each cm^3 of volume contains $6 \cdot 02 \times 10^{20}$ mole, so that the rate of reaction of hydrogen iodide at a concentration of 1 mole per litre is

$$\frac{3 \cdot 4 \times 10^{14}}{6 \cdot 02 \times 10^{20}} = 5 \cdot 5 \times 10^{-7} \text{ 1 mole}^{-1}\text{s}^{-1}$$

which is the value of k, since rate of reaction is $k[HI]^2$ ($[HI] = 1$). This compares favourably with the experimental value of $3 \cdot 5 \times 10^{-7}$ litre mole^{-1} s^{-1} at this temperature.

The collision theory does not give such good agreement for a number of reactions, especially where more complicated mechanisms are involved. An alternative theory, known as the absolute rate theory, is used to calculate, by statistical methods, the frequency with which an activated complex (which subsequently breaks down to give the products) is produced.

Methods of following the progress of a reaction A wide variety of techniques are used to indicate the progress of a reaction. For gaseous reactions involving a volume change, pressure changes can be monitored. Typical examples are the decomposition of cyclobutane

$$C_4H_8 = 2C_2H_4$$

and the decomposition of acetic anhydride

$$(CH_3CO)_2O = CH_3COOH + CH_2CO$$

Reactions which produce an acid can be followed either by titration or by measuring pH values, e.g. the decomposition of tertiary butyl formate:

$$HCOOC(CH_3)_3 = HCOOH + (CH_3)_2C = CH_2$$

and the decomposition of cyclohexyl bromide:

$$C_6H_{11}Br = C_6H_{10} + HBr$$

Conductivity, dilatometry and colorimetry, etc. are other alternative techniques which may be employed in this work.

Catalysis

A catalyst is a substance which increases or decreases the speed of a chemical reaction, but is not consumed by the reaction. A catalyst may either be in the same physical state and be mixed with the reaction mixture (*homogeneous catalysis*), or in a different physical state (*heterogeneous catalysis*). Catalysts which reduce the speed of a reaction are known as *negative catalysts*.

General properties of catalysts
1. They may be poisoned (i.e. their activity impaired or stopped) by a fourth substance.
2. They are specific in their action; a given catalyst will only alter

the speed of a particular reaction, or of reactions of one particular type (such as hydrogenations). In addition, different catalysts may give different end products from the same reactants; for example, formic acid forms carbon dioxide when passed over hot zinc oxide, but carbon monoxide is produced when it is heated in the presence of aluminium oxide.
3. Small quantities of catalyst usually are sufficient to influence the speed of a reaction. For instance, one part of colloidal platinum per million parts of hydrogen peroxide catalyses the decomposition of the latter. In some cases, as in the Friedel Craft reaction, for example, larger quantities of catalyst are needed.
4. A catalyst does not alter the composition of the mixture produced in an equilibrium reaction, but the equilibrium state is attained more rapidly.
5. When the catalyst is a solid, the catalytic effect is usually more marked the finer the state of subdivision of the catalyst.
6. The activity of a catalyst can be increased by the addition of another substance (known as a *catalyst promoter*) which itself shows no catalytic activity.
7. Some reactions, contrary to the law of mass action, speed up as the reaction progresses. This is due to the fact that a product of the reaction acts as a catalyst for the reaction—this is known as *auto-catalysis*.

Mechanism of Catalysis

A catalyst is capable of providing an alternative course, requiring a lower activation energy, for the reaction. Two theories have been proposed to account for the action of catalysts.

Intermediate compound theory A reaction $X + Y \rightarrow XY$ may, in the presence of a catalyst, take place in two stages, the first of which is the formation of an intermediate compound between the catalyst and a reactant. For example,

$$X + C + Y \rightarrow XC + Y$$
$$\downarrow$$
$$XY + C$$

(Here, C represents the catalyst, which is regenerated during the reaction.) Either the isolation of the intermediate compound, or some evidence to indicate that it has been formed, would substantiate this theory. The fact that the physical form of the catalyst may be changed during the reaction agrees with the theory; for example, lumpy manganese dioxide used as a catalyst in the preparation of oxygen from potassium chlorate is reduced to a fine powder.

Surface action theory Where a phase boundary exists between two substances, it is usually found that the concentration of molecules at the interface is greater than the concentration in the bulk of the phase. Figure 61 shows this effect—known as *adsorption*—at a solid–gas interface. A heterogeneous catalyst may function by bringing the reacting molecules into close contact at the surface of the catalyst. As an example, the reaction $C_2H_4 + Br_2 \rightarrow C_2H_4Br_2$ takes place readily at 200°C in a glass vessel, and even more rapidly if the flask contains glass wool or beads, giving extra surface area. However, if the inner surface of the vessel is coated with wax, the reaction comes to a halt.

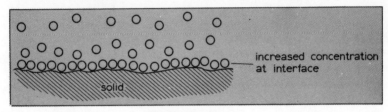

Fig. 61. Adsorption

EXERCISES

1. 23 g of ethanol and 30 g of acetic acid were refluxed until equilibrium was attained. The equilibrium mixture was found to contain 29·3 g of ethyl acetate and 6 g of water. Calculate the weight of ethyl acetate that would be formed at equilibrium by refluxing 46 g of ethanol and 40 g of acetic acid.
2. 2 g of hydrogen iodide was kept at 400°C until equilibrium had been established, when it was found that the iodine present titrated against 30 cm³ of 0·1M sodium thiosulphate solution. Calculate the value of the equilibrium constant for the reaction $2HI \rightleftharpoons H_2 + I_2$.
3. Carbon dioxide dissociates on heating according to the equation

$$2CO_2 \rightleftharpoons 2CO + O_2$$

If, at a certain temperature and one atmosphere pressure, 60% of the original carbon dioxide remains undissociated, calculate K_p for the equilibrium.

4. Calculate the weight of phosphorus pentachloride present at equilibrium when one mole of phosphorus pentachloride is heated to 200°C. The equilibrium constant K_c for the reaction $PCl_5 \rightleftharpoons PCl_3 + Cl_2$ at 200°C is 3·0, and the volume occupied by the system is 2 litres.

5. 32 g of hydrogen and 112 g of nitrogen are sealed with a suitable catalyst in a pressure vessel fitted with a pressure gauge. The temperature was kept constant until equilibrium was established, when it was found that the pressure in the vessel had fallen to 9/10ths of its original value. Calculate,

   (a) the number of moles of hydrogen and nitrogen present initially,
   (b) the total number of moles of gas present initially,
   (c) the total number of moles of gas present at equilibrium,
   (d) the number of moles of hydrogen, nitrogen and ammonia present at equilibrium,
   (e) the value of the equilibrium constant (if the volume of the pressure vessel is V litres) for the reaction $N_2 + 3H_2 \rightleftharpoons 2NH_3$.

6. Calculate the equilibrium constant for the reaction $KI + I_2 \rightleftharpoons KI_3$ from the following results:

   Weight of potassium iodide present initially = 20·75 g
   Weight of free iodine present at equilibrium = 0·0127 g
   Number of moles of KI_3 present at equilibrium = 0·005

7. If N_2O_4 is 36% dissociated at 2 atmospheres pressure and 40°C, what is the value of K_p?

8. If the density of air is 15 times that of hydrogen and the density of phosphorus pentachloride vapour (at 250°C) is four times that of air, calculate the degree of dissociation of phosphorus pentachloride at 250°C.

9. When 500 cm³ of an aqueous solution containing 5 g of an organic substance P was shaken with 100 cm³ of benzene, 2 g of P remained in the aqueous layer. Calculate the weight of P left in the aqueous layer after shaking with another 100 cm³ of benzene, assuming that P is in the same molecular species in each solvent.

10. If the relative (vapour) density of dinitrogen tetroxide is 38·5 at 25°C and 25·0 at 100°C, calculate the degree of dissociation in each case and deduce whether dissociation is an endothermic or exothermic process.

11. The relative density of phosphorus pentachloride is 75 at 160°C. What is the degree of dissociation at this temperature?

12. 50 cm^3 of 0·1M iodine in carbon tetrachloride is shaken with 100 cm^3 of water at 20°C. If 25 cm^3 of the water layer required 5·2 cm^3 of 0·01M sodium thiosulphate to react with the extracted iodine, calculate the value of the distribution coefficient of iodine between carbon tetrachloride and water at 20°C.
13. An aqueous solution contains 10 g of a certain phenol per litre. When 100 cm^3 of this solution is shaken with 20 cm^3 of ether, the ether layer extracts 0·8 g of the phenol. Calculate the volume of ether that would be required to extract 80% of the phenol from 500 cm^3 of the aqueous solution, in one extraction. Also, calculate the weight of phenol that would be obtained by two separate extractions, using half this calculated volume of ether in each extraction.
14. The following figures refer to the distribution of benzoic acid between water and benzene. Calculate the values of the ratios C_1/C_2 and $C_1/\sqrt{C_2}$ and comment on the result.

Concentration of benzoic acid in water (C_1)	1·50	2·77	3·68
Concentration of benzoic acid in benzene (C_2)	2·42	8·24	14·55

15. If the velocity constants for the decomposition of N_2O_5 are 0·30 at 67°C and 0·10 at 57°C, calculate the activation energy for this reaction.
16. The activation energy for the reaction

$$2HI \rightleftharpoons H_2 + I_2$$

is 186 kJ mole^{-1}. Estimate the velocity constant for the reaction at 600 K if k at 560 K is 3.5×10^{-7} litre mole^{-1} s^{-1}. Take R as 8·32 J.

6. Electrochemistry and Ionic Equilibrium

We have already seen that chemical reactions are accompanied by energy changes, and Chapter 4 is devoted to a discussion of the significance of heat energy changes in chemistry. However, other forms of energy are associated with chemical reactions; for example, when electrical energy is applied to a system, a chemical change may result. Alternatively, a chemical reaction may liberate electrical energy. The study of the relation between chemical change and electrical energy is known as *electrochemistry*. Matter is electrical in nature and many reactions involve the transfer of charged chemical units (ions) or electrons. Such exchanges are considered under the heading of *ionic equilibria*.

Electrochemistry

When bars of two dissimilar metals—iron and copper, for example— are placed in a suitable solution such as dilute sulphuric acid, an electric current is detected in the wire connecting the two bars. This system is called a *voltaic* or *galvanic cell*. The electron flow which constitutes the current in the external circuit (i.e. the wire joining the electrodes) is set up under the influence of the electromotive force (e.m.f.) of the cell. The e.m.f. of the cell is equal to the potential difference (in volts) between the two electrodes. Any conductor used in the external circuit connecting the two electrodes has the property of resisting (to a larger or smaller extent) the electron flow. This property is termed *resistance* and it is measured in ohms.

Ohm's law *The current (I) flowing in the external circuit is directly proportional to the e.m.f. (E) and is inversely proportional to the resistance (R).*

That is,

$$I = \frac{E}{R}$$

where I is given in amperes, E is in volts and R is in ohms. Quantity of electricity is measured as the product of the current strength and time. The unit is the *coulomb*, which is defined as the quantity of electricity passing when a current of one ampere flows for one second. Quantity of electricity may be determined experimentally from the weight of metal deposited in a coulometer (an electrolytic cell designed for accurate measurements of this type). When electrical energy is converted into heat energy (i.e. by measuring the heat produced in a resistance wire) it is found that the heat produced is proportional to the e.m.f. in the circuit and the quantity of electricity passing, i.e.

$$\text{electrical energy} = EIt$$

(t = time in seconds). The unit is the volt-coulomb or *joule*.

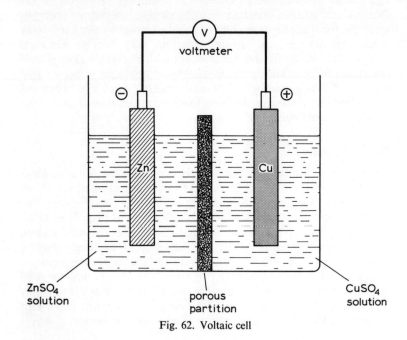

Fig. 62. Voltaic cell

Voltaic cells Figure 62 shows a typical voltaic cell (a Daniell cell) in which a rod of zinc dips into a solution of zinc sulphate and a rod of copper dips into a solution of copper sulphate. The two solutions are separated by a porous partition. A potential difference exists between the two electrodes, and a current flows in the external circuit. This current is produced by the dissolution of the zinc electrode by the reaction:

$$\underset{\text{electrode}}{\text{Zn}} \rightarrow \underset{\substack{\text{in} \\ \text{solution}}}{\text{Zn}^{2+}} + \underset{\substack{\text{flow from electrode to} \\ \text{external circuit}}}{2e^-} \quad \text{(a)}$$

and deposition of copper by the reaction:

$$\underset{\text{in solution}}{Cu^{2+}} + \underset{\substack{\text{from} \\ \text{external} \\ \text{circuit}}}{2e^-} \rightarrow \underset{\text{deposited on electrode}}{Cu} \qquad (b)$$

The net effect is that two electrons (per atom of metal) have been transferred, via the external circuit, from the zinc metal to the copper ions. As a result, zinc has been oxidized while copper (II) ions have been reduced. This is in accordance with the definition (see Part 1 of this series) that reduction corresponds to the gain of electrons and the loss of electrons corresponds to oxidation. The overall equation is found by linking equations (a) and (b) for the two electrode reactions:

$$Zn + Cu^{2+} = Zn^{2+} + Cu$$

Equations such as (a) or (b) are known as *half equations* or *half reactions*.

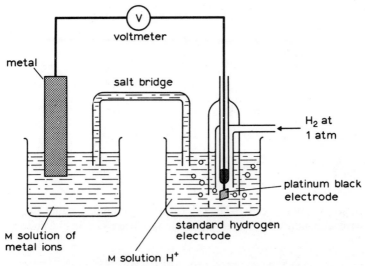

Fig. 63. Measurement of standard electrode potentials

Standard electrode potentials In the Daniell cell, the zinc loses electrons (oxidizes) with a driving force which exceeds the resistance to acceptance of electrons by the copper ions (reduction). The driving force with which metals lose electrons is measured as a *standard electrode potential*, which is defined as the potential difference between a standard hydrogen electrode and a metal which is immersed in a solution containing metal ions at molar concentration at 25°C. The apparatus used for this determination is shown diagrammatically in Figure 63.

A standard hydrogen electrode is formed by bubbling hydrogen gas at one atmosphere pressure round a platinum electrode coated with platinum black, which is in contact with a molar solution of hydrogen ion at 25°C. The platinum black catalyses the establishment of the equilibrium

$$\tfrac{1}{2}H_2 \rightleftharpoons H^+ + e^-$$

The electrodes are connected through a voltmeter, and the solutions are joined by means of a salt bridge (a tube packed with potassium chloride slurry). The standard electrode potentials (designated E^0) of zinc and copper are $-0\cdot76$ V and $+0\cdot34$ V respectively—the potential of the standard hydrogen electrode being taken as zero for reference purposes. A list of E^0 values for some of the commoner metals is given below:

Half reaction	E^0
$K^+ + e^- = K$	$-2\cdot92$ V
$Ca^{2+} + 2e^- = Ca$	$-2\cdot87$ V
$Na^+ + e^- = Na$	$-2\cdot71$ V
$Mg^{2+} + 2e^- = Mg$	$-2\cdot37$ V
$Al^{3+} + 3e^- = Al$	$-1\cdot67$ V
$Zn^{2+} + 2e^- = Zn$	$-0\cdot76$ V
$Fe^{2+} + 2e^- = Fe$	$-0\cdot44$ V
$Pb^{2+} + 2e^- = Pb$	$-0\cdot13$ V
$H^+ + e^- = \tfrac{1}{2}H_2$	$0\cdot00$ V
$Cu^{2+} + 2e^- = Cu$	$+0\cdot34$ V
$Ag^+ + e^- = Ag$	$+0\cdot80$ V

Representation of cells The cell formed by coupling a zinc electrode with a standard hydrogen electrode is represented in the following way:

```
                  ........→.....         flow of electrons in external circuit

        Zn    |  Zn²⁺    :  H⁺     |  H₂/Pt
     Electrode|  Molar   :  Molar  |  Electrode
              |  solution:  solution|
        −            SALT         +
                    BRIDGE
```

The IUPAC convention requires that the positive electrode should

appear on the right of this cell diagram, and the e.m.f. of the cell is given by
$$E = E_{right} - E_{left}$$
(E_{right} is the potential of the right hand electrode, etc.)

Similarly, for the cell formed by coupling the copper and standard hydrogen electrodes, we have:

$$\text{Pt/H}_2 \left| \begin{array}{c} \text{H}^+ \\ [\text{H}^+]=1 \\ - \end{array} \right. \vdots \left. \begin{array}{c} \text{Cu}^{2+} \\ [\text{Cu}^{2+}]=1 \end{array} \right| \begin{array}{c} \text{Cu} \\ \\ + \end{array} \qquad \text{I}$$

Electrode reactions $H_2 \rightarrow 2H^+ + 2e^-$ $Cu^{2+} + 2e^- \rightarrow Cu$

Therefore, we may represent the Daniell cell by:

$$\text{Zn} \left| \begin{array}{c} \text{Zn}^{2+} \\ [\text{Zn}^{2+}]=1 \\ - \end{array} \right. \vdots \left. \begin{array}{c} \text{Cu}^{2+} \\ [\text{Cu}^{2+}]=1 \end{array} \right| \begin{array}{c} \text{Cu} \\ \\ + \end{array}$$

Electrode reactions $Zn \rightarrow Zn^{2+} + 2e^-$ $Cu^{2+} + 2e^- \rightarrow Cu$
and
$$\begin{aligned} E &= E_{right} - E_{left} \\ &= 0.34 - (-0.76) = 1.10 \text{ V} \end{aligned}$$

The e.m.f. of a cell (for standard conditions of temperature and electrolyte concentration) is found by subtracting the more negative E^0 value from the more positive. For example, the e.m.f. of the cell formed by coupling an iron and a copper electrode would be

$$\begin{aligned} E &= E^0_{copper} - E^0_{iron} \\ &= 0.34 - (-0.44) = 0.78 \text{ V} \end{aligned}$$

Redox potentials In the cell diagram I above, electrons are given to copper ions (at the copper electrode) which are thereby reduced to the metal. The electrode takes no part in the reaction other than feeding electrons to the ions in solution. The same effect may be produced if we replace the copper electrode by an inert electrode (such as platinum) which dips into a solution containing a species which may be reduced to some lower oxidation state (see Part 1) rather than to the metal itself. A solution containing iron (III) and iron (II) ions is an example:

$$Fe^{3+} + e^- \rightleftharpoons Fe^{2+}$$

When coupled with a standard hydrogen electrode, this system may either

(a) accept electrons from the external circuit corresponding to the reaction

$$Fe^{3+} + e^- \rightarrow Fe^{2+}$$

in which case, the electrode will appear positive with respect to the hydrogen electrode, and iron (III) is reduced to iron (II), or

(b) feed electrons, by way of the inert electrode, into the external circuit, corresponding to the reaction

$$Fe^{2+} \rightarrow Fe^{3+} + e^-$$

In this case, the electrode will appear negative with respect to the standard hydrogen electrode, and iron (II) is oxidized.

The value found by experiment is 0·76 V showing that alternative (a) is favoured. In order to compare the oxidizing or reducing powers of half-reactions, standard conditions must obtain. These are specified as:

(a) The concentration of the oxidized form must equal the concentration of the reduced form, and both should be molar.
(b) The pH of the solution must be zero.
(c) The temperature should be 25°C.

The standard potentials measured in this way are termed reduction–oxidation (or *redox*) potentials, and a list of these for some common half-reactions follows:

Redox Potentials for Common Half-Reactions

Oxidized form	Reduced form	$E°$ (volt)	
$K^+ + e^-$	$= K$	$-2·92$	
$Ca^{2+} + 2e^-$	$= Ca$	$-2·86$	
$Na^+ + e^-$	$= Na$	$-2·71$	↑
$Mg^{2+} + 2e^-$	$= Mg$	$-2·37$	
$Al^{3+} + 3e^-$	$= Al$	$-1·67$	Reduced form
$Se + 2e^-$	$= Se^{2-}$	$-0·77$	increases in
$Zn^{2+} + 2e^-$	$= Zn$	$-0·76$	reducing power
$S + 2e^-$	$= S^{2-}$	$-0·51$	
$Fe^{2+} + 2e^-$	$= Fe$	$-0·44$	
$Sn^{2+} + 2e^-$	$= Sn$	$-0·14$	
$Pb^{2+} + 2e^-$	$= Pb$	$-0·13$	
$2H^+ + 2e^-$	$= H_2$	zero	
$Cu^{2+} + 2e^-$	$= Cu$	$0·35$	
$I_2 + 2e^-$	$= 2I^-$	$0·58$	
$O_2 + 2H^+ + 2e^-$	$= H_2O_2$	$0·68$	
$Fe^{3+} + e^-$	$= Fe^{2+}$	$0·76$	
$Br_2 + 2e^-$	$= 2Br^-$	$1·07$	Oxidized form
$Cr_2O_7^{2-} + 14H^+ + 6e^-$	$= 2Cr^{3+} + 7H_2O$	$1·33$	increases in
$Cl_2 + 2e^-$	$= 2Cl^-$	$1·36$	oxidizing power
$MnO_4^- + 8H^+ + 5e^-$	$= Mn^{2+} + 4H_2O$	$1·52$	
$F_2 + 2e^-$	$= 2F^-$	$2·83$	↓

For practical purposes, a hydrogen electrode is not very convenient to work with, and an alternative, such as a calomel electrode, is employed. A calomel electrode, shown in Figure 64, has a potential of +0·334 V with respect to hydrogen, and this must be added to the potential difference between a test electrode and a calomel electrode in order to obtain the E^0 value. For example, if the p.d. between an iron and a calomel electrode under standard conditions is −0·776 V, the E^0 value for the half-reaction

$$Fe^{2+} + 2e^- = Fe$$

is

$$-0·776 + 0·334 = -0·442 \text{ V}$$

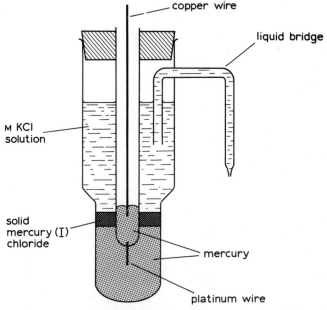

Fig. 64. Calomel electrode

Uses of redox potentials

(a) *Oxidizing and reducing power* If a half-reaction tends to change from the reduced form to the oxidized form, electrons are liberated, and a negative potential develops. Therefore, the most powerful reducing agents are the *reduced species* in the half-reactions having the highest negative E^0 values (i.e. potassium, calcium and sodium). Conversely, a system which has good oxidizing properties changes from the oxidized to the reduced form and accepts electrons. Such systems have positive E^0 values, and the strongest oxidizing agents are the *oxidized forms* of the half-reactions having the largest positive E^0 voltages. A list of redox potentials is a relative sequence, hence the fact that a half-reaction has a positive E^0 voltage does not automatically mean that it must behave as an oxidizing agent. For example, iron (III) oxidizes iodide ion to iodine, but chlorine oxidizes iron (II) to iron (III). These conclusions can only be arrived at when two half-reactions are coupled.

(b) *Probability of reaction* A redox reaction is possible when an oxidized and a reduced form are mixed. To deduce whether or not such a mixture can react, we subtract one half-reaction from the other as follows:

$$\tfrac{1}{2}Cl_2 + e^- = Cl^- \qquad E_1^0 = 1.36 \text{ V} \qquad (1)$$
$$Fe^{3+} + e^- = Fe^{2+} \qquad E_2^0 = 0.76 \text{ V} \qquad (2)$$

Subtracting (2) from (1) we get

$$Fe^{2+} + \tfrac{1}{2}Cl_2 = Fe^{3+} + Cl^-$$
$$E^0 = E_1^0 - E_2^0$$
$$= 1.36 - 0.76 = 0.60 \text{ V}$$

Now it may be shown that

$$\Delta G^0 = -nFE^0$$

where ΔG^0 is the standard free energy change for the reaction and n is equal to the number of electrons transferred in the reaction, F is the Faraday constant (96 500 coulombs per mole) and E^0 is the standard redox potential for the reaction. In this case,

$$\Delta G^0 = -1 \times 96\,500 \times 0.6 = -57\,900 \text{ J mole}^{-1}$$

Without calculating the actual value of ΔG^0, we can see that a positive E^0 value (which makes ΔG^0 negative) for the reaction means that the reaction (going from left to right) is thermodynamically feasible. As always, the rate of reaction depends on kinetic factors.

Example 61 Given that

$$Cr^{3+} + e^- = Cr^{2+} \qquad\qquad E_1^0 = -0.41 \text{ V} \quad (1)$$
$$H_3PO_4 + 2H^+ + 2e^- = H_3PO_3 + H_2O \qquad E_2^0 = -0.28 \text{ V} \quad (2)$$

deduce whether Cr^{3+} can be reduced by phosphorous acid under standard conditions.

Solution In order to subtract (2) from (1), equation (1) must be multiplied by two so that the electron supply and demand balances. Now E^0 values are (like temperature) independent of the amount of material present. Therefore, if we multiply the quantities in the equation to balance the stoichiometry, we *do not* multiply the E^0 values. Hence

$$2Cr^{3+} + 2e^- = 2Cr^{2+} \qquad E_1^0 = -0.41 \text{ V}$$
$$H_3PO_4 + 2H^+ + 2e^- = H_3PO_3 + H_2O \qquad E_2^0 = -0.28 \text{ V}$$

Subtracting,

$$2Cr^{3+} + H_3PO_3 + H_2O = 2Cr^{2+} + H_3PO_4 + 2H^+$$
$$E^0 = E_1^0 - E_2^0$$
$$= -0.41 - (-0.28) = -0.13 \text{ V}$$

This means that ΔG^0 for the reaction going from left to right is positive, and we would conclude that it is not possible to reduce Cr^{3+} by phosphorous acid under standard conditions.

(c) *Disproportionation*   Taking the two half-reactions for copper:

$$Cu^+ + e^- = Cu \qquad E_1^0 = 0.5 \text{ V} \qquad (1)$$
$$Cu^{2+} + e^- = Cu^+ \qquad E_2^0 = 0.1 \text{ V} \qquad (2)$$

and subtracting equation (2) from equation (1) we get

$$2Cu^+ = Cu + Cu^+$$
$$E^0 = E_1^0 - E_2^0 = 0.4 \text{ V}$$

This gives ΔG a negative value, and reaction from left to right is thermodynamically feasible. This type of reaction is an example of disproportionation, in which a species undergoes simultaneous self-oxidation and reduction. Disproportionation is feasible if the E^0 value for the reduction half-reaction is more positive than that for the oxidation. Further applications of redox potentials in inorganic chemistry are mentioned in Part 1 of the series (General and Inorganic Chemistry).

Redox potentials in non-standard states   The E^0 value for a half-reaction (i.e. for a cell in which the potential of the test electrode is compared against a standard hydrogen electrode) is measured when:

(a) the concentration of both oxidized and reduced forms in the electrolyte in the test cell is molar,
(b) the pH is zero,
(c) the temperature is 25°C (298 K).

When the temperature, pH or concentrations deviate from the standard conditions, the redox potential (E) differs from the standard value (E^0). The relation between E and E^0 is given by the Nernst equation:

$$E = E^0 + \frac{R \times T \times 2.303}{n \times F} \log \frac{\text{[oxidized form]}}{\text{[reduced form]}}$$

(where n equals the number of electrons transferred).

Example 62 The E^0 value for the half-reaction

$$Fe^{3+} + e^- = Fe^{2+}$$

is 0·758 V. What is the redox potential at 50°C if the concentrations of Fe^{3+} and Fe^{2+} are 0·1M and 1M respectively?

Solution Assuming the pH to be zero, and using the Nernst equation,

$$E = 0.758 + \frac{8.32 \times 323 \times 2.303}{1 \times 96\,500} \log \left(\frac{0.1}{1}\right)$$

$$= 0.758 + 0.064 \log 10^{-1} = 0.694 \text{ V}$$

The redox potentials corresponding to metal ions in two oxidation states (e.g. Fe^{3+}/Fe^{2+}, Co^{3+}/Co^{2+}) are often significantly changed in the presence of a complexing agent. This is due to the fact that a stronger complex is formed between the ligands and the metal ion in the higher oxidation state (oxidized form). Hence, the effective concentration* of the oxidized form is greatly reduced, which can be seen from the Nernst equation to give an E value which is less than E^0. For instance, E^0 for Fe^{3+}/Fe^{2+} is 0·76 V, but in the presence of cyanide ion this may fall to 0·5 V.

Several half-reactions involve the presence of hydrogen ion. For example,

$$MnO_4^- + 8H^+ + 5e^- = Mn^{2+} + 4H_2O \qquad E^0 = 1.52 \text{ V}$$

The oxidized form of this half-reaction includes hydrogen ion, and the E values are pH dependent. The Nernst equation takes the form:

$$E = E^0 + \frac{2.303\,RT}{nF} \log \frac{[MnO_4^-][H^+]^8}{[Mn^{2+}]}$$

As the pH rises, $[H^+]$ becomes less than one, and the logarithm term takes on a negative value. This has the effect of making E steadily less than E^0 as the pH rises—that is, permanganate ion loses

* The effective concentration of a species is given in terms of activities (p. 203).

some of its oxidizing power as the pH increases. This agrees with the fact that at pH 6, (when $E = 0.94$ V) iodide ion is the only halide ion to be oxidized by permanganate. This factor also shows why the preparation of some powerful oxidizing agents such as perchlorate, permanganate and dichromate is more easily accomplished under alkaline conditions. No hydrogen ions appear in the half-reaction

$$Cl_2 + 2e^- = 2Cl^- \qquad E^0 = 1.36 \text{ V}$$

and changes in pH do not alter E^0. This is not necessarily true for all half-reactions of this type, as pH changes can affect the system in a more indirect manner.

Concentration cells If we apply the Nernst equation to a standard electrode potential in which the reduced form is the pure metal (M) we get

$$E = E^0 + \frac{2 \cdot 303 RT}{nF} \log \frac{[M^{n+}]}{[M]}$$

which reduces to

$$E = E^0 + \frac{2 \cdot 303 RT}{nF} \log[M^{n+}]$$

since the effective concentration (activity—see p. 204) of a pure solid is taken as unity. If a cell is set up by connecting two electrodes of the same metal dipping into metal ion solutions of different concentrations, a small potential difference will be set up, and such a cell will produce a small e.m.f. This arrangement is known as a concentration cell. For example:

$$\text{Cu} \ \bigg| \ \begin{matrix} Cu^{2+} \\ (M/100) \end{matrix} \ \bigg\vdots \ \begin{matrix} Cu^{2+} \\ (M) \end{matrix} \ \bigg| \ \text{Cu}$$

E of the left-hand electrode can be calculated from the Nernst equation to be 0·28 V at 25°C. For the electrode on the right, $E = E^0 = 0.34$ V. This electrode is positive, so we can say

$$\begin{aligned} E &= E_{\text{right}} - E_{\text{left}} \\ &= 0.34 - 0.28 = 0.06 \text{ V} \end{aligned}$$

Electrodes used to measure pH The potential of a hydrogen electrode depends on the concentration of the hydrogen ions surrounding the electrode.

$$H^+ + e^- = \tfrac{1}{2}H_2$$

$$E = E^0 = \frac{2\cdot 303 RT}{F}\log[H^+]$$

$$= \frac{-2\cdot 303 RT}{F}(\text{pH})$$

since $E^0 = 0$, $n = 1$ and $[H_2]$ is taken as unity. Hence, comparison of the potential difference set up between a hydrogen electrode and another electrode whose E value is independent of pH forms the basis of an electrode system which can be used to measure the pH of a solution. A hydrogen electrode is difficult to use in practice, and an alternative electrode (such as a glass electrode) whose E value also depends on pH is substituted. This system, in instrument form, functions as a pH meter.

Fuel cells In a voltaic cell, the free energy of a chemical reaction is converted into electrical energy. A cell that utilizes the combustion of a fuel as the chemical reaction which provides the electrical energy is called a *fuel cell*. Most fuel cells use hydrogen and oxygen as reactants, although hydrocarbon and alcohols can take the place of hydrogen. None of these fuels react with oxygen at room temperature, and a catalyst has to be employed to increase the rate of reaction. A diagram of a hydrogen–oxygen fuel cell is shown in Figure 65. The gases pass under pressure round the anodes and cathodes (made from porous nickel) immersed in potassium hydroxide solution.

At the cathode:

$$\tfrac{1}{2}O_2 + H_2O + 2e^- = 2OH^-$$

At the anode:

$$H_2 + 2OH^- = 2H_2O + 2e^-$$

Overall:

$$H_2 + \tfrac{1}{2}O_2 = H_2O$$

for which

$$\Delta G = -238 \text{ kJ mole}^{-1}$$

and, using $\Delta G = -nFE$,

$$E = \frac{238\,000}{2 \times 96\,500} = 1\cdot 23 \text{ V}$$

The production of electricity by means of a fuel cell is reported to be twice as efficient as the method in which the fuel is burnt to power a steam turbine.

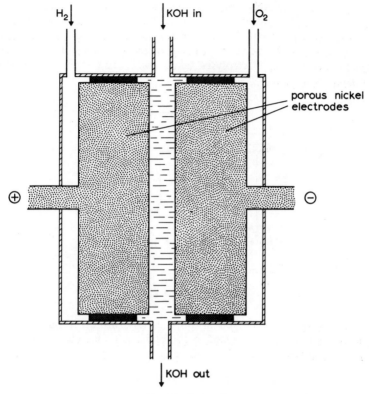

Fig. 65. Fuel cell

Electrolysis

An electric current may be conducted by:
  (a) metals, in which case no chemical change is noticed, although a physical effect (e.g. heating and magnetic effects) may result. This is *metallic conduction*.
  (b) solutions of electrolytes. These solutions conduct an electric current and are decomposed by it. This type of conduction is *electrolytic* conduction.

In Figure 66, the current is led into and out of solution by electrodes—the positive electrode is the *anode* and the negative electrode is the *cathode*. A flow of electrons, forced round the circuit by the battery, constitutes the current; the cathode supplies electrons to the ions in the solution, while the anode accepts electrons from the ions in solution. Positively charged ions (*cations*) are attracted to the cathode, while negatively charged ions (*anions*) travel towards the anode. As the current flows, electrons are lost by the cathode and are released to the anode as a result of the discharge of ions; thus decomposition or electrolysis takes place. The whole assembly constitutes an *electrolytic cell*.

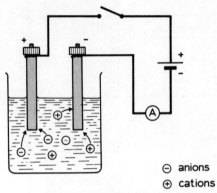

⊖ anions
⊕ cations

Fig. 66. Electrolytic cell

The current is measured in amperes (A), while the quantity of electricity passed is measured in coulombs (C). One coulomb is the quantity of electricity passed when one ampere flows for one second:

$$\text{quantity of electricity passed} = I \times t$$

where I is the current in amperes and t the time in seconds.

Faraday's Laws of Electrolysis

Quantitative investigations of electrolysis were carried out by Faraday, and the results of this work are summarized in Faraday's laws of electrolysis, published in 1834.

The *first law* states that *the weight of an individual substance liberated during electrolysis is proportional to the quantity of electricity which passes through the electrolyte.* That is,

$$\text{weight liberated} \propto I \times t$$

This law is concerned with the passage of different quantities of electricity through the same electrolyte.

The *second law* states that *the weights of various substances liberated* by the passage of the same quantity of electricity through different electrolytes are proportional to the chemical equivalent of the ion. That is,

$$\text{weight liberated} \propto E$$

where E is an equivalent weight of the ion, given by the ratio

$$E = \frac{\text{weight of 1 mole of ion}}{\text{charge carried by that ion}}$$

Example 63 Calculate the equivalent weights of (a) chloride ion, (b) barium ion, (c) sulphate ion.

Solution The ions are charged as follows: Cl^-, Ba^{2+}, SO_4^{2-}. Hence, the values of E for these ions are:

$$\frac{35\cdot46}{1} = 35\cdot46 \quad \text{for } Cl^-$$

$$\frac{137\cdot4}{2} = 68\cdot7 \quad \text{for } Ba^{2+}$$

$$\frac{96\cdot06}{2} = 48\cdot03 \quad \text{for } (SO_4)^{2-}$$

Combining the two laws of electrolysis, we get

$$\text{weight liberated} \propto EIt$$

or

$$\text{weight liberated} = kEIt$$

By experiment, k is found to be $1/96\,500$; i.e., $96\,500$ coulombs of electricity will liberate one mole of univalent ion, or half a mole of divalent ion, etc. This quantity of electricity is called the Faraday constant (F).

Thus,

$$\text{weight liberated } W = \frac{EIt}{F} = \frac{EIt}{96\,500}$$

$$= zIt$$

where z is known as the *electrochemical equivalent* of the ion and has the value $E/96\,500$.

Example 64 If a current of 2 A is passed for 1 h 35 min through dilute sulphuric acid, what volume of hydrogen is liberated at 20°C and 750 mm pressure?

Solution The hydrogen ion carries a single positive charge. Hence,

$$W = \frac{EIt}{96\,500} = \frac{1\cdot008 \times 2 \times 5700}{96\,500}$$
$$= 0\cdot118 \text{ g}$$

Now one mole of hydrogen (2·016 g) occupies 22·4 litres at s.t.p
Therefore 0·118 g occupies

$$\frac{22\cdot4 \times 0\cdot118}{2\cdot016} \text{ litres at s.t.p.}$$

or

$$\frac{22\,400 \times 0\cdot118 \times 760 \times 293}{2\cdot016 \times 273 \times 750} \text{ cm}^3 \text{ under the given conditions}$$
$$= 10\,200 \text{ cm}^3$$

The discharge of ions If we measure the current flowing through an electrolytic cell containing, say, a molar solution of hydrochloric acid, a graph of the current against the applied voltage may be obtained (Figure 67). Since an electrolysis is the reverse of a voltaic

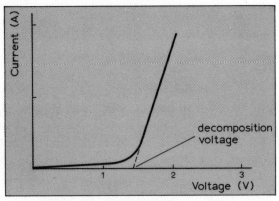

Fig. 67

cell, no current should flow in the circuit until the applied e.m.f. exceeds that produced by coupling the two electrodes. Under standard conditions this value is zero (E^0 for H^+/H_2 Pt) plus 1·36 V (E^0 for $Cl_2/2Cl^-$) for this particular case. When this voltage (called the *decomposition voltage*) is exceeded, electrolysis takes place and ions are discharged at the electrodes. It often happens that the observed decomposition voltage is above the value calculated from standard potentials. This excess is referred to as *overvoltage*, and both electrode processes contribute individually to the total overvoltage for the cell. Overvoltage is a consequence of the high activation energy needed to discharge ions in the form of their final products at an electrode. The value of the overvoltage depends, among other things, on the nature of the electrode; hydrogen, for example, has a large overvoltage at a mercury surface, and oxygen has a high overvoltage at a platinum electrode.

Examples of electrolyses Re-writing the list of redox potentials given on p. 182 gives an indication of the ease of discharge of ions at the electrodes during electrolysis.

Na^+	-2.7 V	Increasing	F^-	2.9 V
Al^{3+}	-1.7 V	ease of	SO_4^{2-}	
Zn^{2+}	-0.8 V	discharge	NO_3^-	
Fe^{2+}	-0.4 V		Cl^-	1.4 V
H^+	0.0 V		Br^-	1.1 V
Cu^{2+}	$+0.3$ V		I^-	0.5 V
Ag^+	$+0.8$ V	↓	OH^-	0.4 V
Ions discharged at cathode			Ions discharged at anode	

Although the calculated decomposition potential for the liberation of hydrogen and oxygen from water is 1·24 V, the overvoltage factor raises this to a higher value (1·8 V, at least, in the case of the electrolysis of very dilute sulphuric acid). Consequently, the electrolysis of aqueous solutions frequently gives products other than hydrogen and oxygen.

The electrolysis of sodium chloride solution At the cathode, hydrogen is discharged by the reaction

$$H_2O + e^- = \tfrac{1}{2}H_2 + OH^-$$

and the electrolyte surrounding the cathode shows an increase in hydroxide ion concentration. When mercury is used as a cathode, the large overvoltage developed by hydrogen at a mercury surface leads to the preferential discharge of sodium ions.

$$Na^+ + e^- = Na$$

The liberated sodium forms an amalgam with the mercury cathode, and this forms the basis of the electrolytic production of sodium hydroxide (see Part 1). At the anode, we would expect hydroxide ion to be discharged since it has a lower decomposition voltage than the chloride ion. Provided:

(a) the chloride concentration is kept high,

(b) the anode material is carbon, which leads to a high overvoltage for oxygen (the final product of the discharge of hydroxide ion), chloride ion is discharged preferentially.

$$Cl^- = \tfrac{1}{2}Cl_2 + e^-$$

If the electrolyte is well stirred, the chlorine reacts with the hydroxide ion to produce sodium hypochlorite at low temperatures,

$$Cl_2 + 2OH^- = Cl^- + OCl^- + H_2O$$

and sodium chlorate above 80°C:

$$3Cl_2 + 6OH^- = 5Cl^- + ClO_3^- + 3H_2O$$

Electrolysis of copper sulphate solution When inert electrodes (e.g. platinum) are used, copper is deposited at the cathode and oxygen is liberated at the anode.

At the cathode:

$$Cu^{2+} + 2e^- = Cu$$

At the anode, hydroxide ion has the lower decomposition voltage:

$$4OH^- = 2H_2O + O_2 + 4e^-$$

If a soluble (sometimes called reactive) anode such as copper, is used, hydroxide ion is not discharged but the anode itself goes into solution.

$$\underset{\text{anode}}{Cu} = \underset{\text{in solution}}{Cu^{2+}} + 2e^-$$

As a result of this dissolution, the anode functions in the usual way in accepting electrons from the dissolving metal. As before, copper is deposited at the cathode, and this process forms the basis of the electrolytic technique for copper refining.

The Conductivity of Electrolyte Solutions

Although metals are classed as good conductors of electricity, they do offer some resistance to the flow of an electric current. To compare the resistances offered by different metals, their *specific resistances* (or *resistivities*) are quoted. The specific resistance of a metal is the resistance of a block 1 cm long and 1 cm^2 in area of cross-section. Solutions of electrolytes have a much higher resistance than metallic conductors, so their *conductivities* are compared, rather than their resistivities. Conductivity is the reciprocal of resistance, and the *specific conductivity** (κ) of a solution is defined as *the reciprocal of the resistance (at a given temperature) of a solution measured between electrodes which are 1 cm² in area of cross-section and placed 1 cm apart.* Since the units of specific resistance are ohm cm, it follows that the units of specific conductivity are the reciprocal of these, that is, ohm^{-1} cm^{-1}.

Measurement of the Specific Conductivity of a Solution

A modified Wheatstone bridge circuit is used (Figure 68), using alternating current (which eliminates polarization effects at the electrodes), and headphones, an oscilloscope or a magic eye device to indicate the balance point.

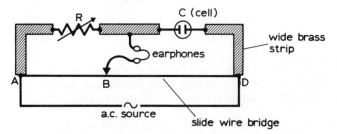

Fig. 68. Measurement of the resistance of an electrolyte

With the conductivity cell in position, a suitable value of R (the variable resistance) is chosen such that the balance point (as indicated by the detector) is towards the centre of the slide wire bridge. At this point, if the resistance of the solution is X, then

$$\frac{X}{R} = \frac{BD}{AB}$$

A variable capacitance is sometimes used in conjunction with the variable resistance in order to balance out capacitance effects of the cell.

A typical conductivity cell is shown in Figure 69. The cell is made from glass or silica, while the electrodes are constructed from stout platinum foil. Very pure de-ionized or distilled water, which has a specific conductivity of not more than 10^{-6} ohm^{-1} cm^{-1}, is used in making up the solutions. Since it is both difficult and time-consuming (and, therefore, expensive) to construct a cell with electrodes which are exactly 1 cm² in area and exactly 1 cm apart at

---

* The term 'electrolytic conductivity' is an alternative designation for this quantity.

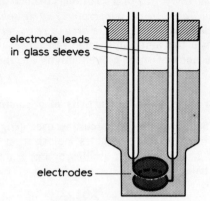

Fig. 69. Conductivity cell

all points, the cell used must have its *cell constant* (which is a multiplying factor to correct for the deviation from the defined dimensions of the electrode system) determined before use. The determination is carried out in the following stages:

1. The cell is thoroughly cleaned, steamed out and dried. It is then filled with exactly M/50 potassium chloride solution.
2. The resistance of the solution is measured and the temperature is noted.
3. By reference to the tabulated data, the specific conductivity of M/50 potassium chloride solution at the observed temperature is ascertained, and the cell constant for the particular cell is calculated, using the relation:

$$\frac{\text{specific conductivity of}}{\text{M/50 potassium chloride}} = \frac{1}{\text{measured resistance}} \times \text{cell constant}$$

4. The cell is cleaned out again, dried, and filled with the solution under test.
5. The resistance of this solution at the given temperature is determined and the specific conductivity of the solution is calculated.

Example 65 At 25°C, the resistance of a solution was 328·5 ohm. The resistance of M/50 potassium chloride solution at the same temperature was 402·3 ohm, and the tabulated specific conductivity of M/50 potassium chloride solution is $0{\cdot}002768\,\text{ohm}^{-1}\,\text{cm}^{-1}$. Calculate the cell constant of the cell used and the specific conductivity of the solution.

Solution

$$\text{Specific conductivity of M/50 potassium chloride} = \frac{1}{\text{measured resistance}} \times \text{cell constant}$$

Thus,
$$\text{cell constant} = 0{\cdot}002\,768 \times 402{\cdot}3 = 1{\cdot}114$$
and hence,
$$\text{specific conductivity of solution} = \frac{1{\cdot}114}{328{\cdot}5} = 0{\cdot}003\,39 \text{ ohm}^{-1}\,\text{cm}^{-1}$$

Since the current is carried in solution by the movement of anions and cations, it follows that the specific conductivity of a solution increases as the concentration of the solution increases.

Equivalent Conductivity*

To compare the relative conductivities of different electrolytes, the solutions used must contain equal quantities of current-carrying species (ions). Thus, a molar solution of sodium chloride, an M/2 solution of barium chloride and an M/3 solution of lanthanum chloride ($LaCl_3$) are equivalent solutions, since they all contain the equivalent of one mole of positive and negative ions. (This corresponds with the definition of an equivalent given on p. 191.)

The equivalent conductivity Λ is related to the specific conductivity by the equation:
$$\Lambda = \kappa V$$
where V is the volume in cm^3 which contains one gramme equivalent.

Example 66 The resistance of an M/10 solution of acetic acid is $2{\cdot}5 \times 10^3$ ohm, when measured in a cell whose cell constant is $1{\cdot}150$. Calculate the equivalent conductivity of M/10 acetic acid.

Solution

$$\text{Specific conductivity} = \frac{1}{\text{measured resistance}} \times \text{cell constant}$$

$$= \frac{1}{2{\cdot}5 \times 10^3} \times 1{\cdot}15$$

$$= 4{\cdot}6 \times 10^{-4} \text{ ohm}^{-1}\,\text{cm}^{-1}$$

* An alternative approach is to use the *molar conductivity*, which is the conductivity of a solution containing 1 mole of electrolyte per litre. It has the value K/C where C is the concentration of the solution in mole m^{-3}.

$$\Lambda = \kappa V \quad (V = 10\,000 \text{ cm}^3)$$
$$= 4 \cdot 6 \times 10^{-4} \times 10\,000$$
$$= 4 \cdot 6 \text{ ohm}^{-1} \text{ cm}^2$$

The Variation of Equivalent Conductivity with Dilution

The term dilution refers to the number of litres of solution which contain one gramme equivalent of electrolyte. When values of the equivalent conductivity of an electrolyte are measured at different dilutions, the results obtained correspond, in general, with one or other of the two types of curve shown in Figure 70. The upper curve

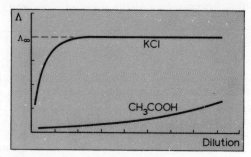

Fig. 70. Variation of equivalent conductivity with dilution

is for potassium chloride, and the lower for acetic acid; the actual values of the equivalent conductivity, in $\text{ohm}^{-1} \text{ cm}^{-2}$, at the various dilutions are:

Dilution	1	10	100	1000	5000	10000	∞
Λ_{KCl}	98·3	112	122	127	128	129	130
Λ_{CH_3COOH}	1·3	4·6	14·3	41	50	107	350

Provided the number of ions formed by the electrolyte remains constant, a simple calculation will show that the value of the equivalent conductivity of a solution should be independent of dilution. Suppose a molar solution of an electrolyte which forms a singly charged cation and a singly charged anion (sodium chloride, for example) contains n positive and n negative ions per cm^3. Let us further assume that this solution has a specific conductivity of $0 \cdot 1 \text{ ohm}^{-1} \text{ cm}^{-1}$, so that the equivalent conductivity is $0 \cdot 1 \times 1000$ or $100 \text{ ohm}^{-1} \text{ cm}^2$. If this solution is diluted so that it is now M/2, there will be $n/2$ positive ions and $n/2$ negative ions per cm^3, and the specific conductivity will fall to one-half of the original value;

that is, it will be now 0·05 ohm^{-1} cm^{-1}. But the equivalent conductivity will be 0·05 × 2000 (since this is the volume which now contains one gramme equivalent) or 100 ohm^{-1} cm^2.

Taking the curve for potassium chloride, the equivalent conductivity has a fairly high value even in a concentrated solution; it increases rapidly at first and then levels off at a maximum value of 130 ohm^{-1} cm^2. This maximum value is obtained at high dilution and is known as the equivalent conductivity at infinite dilution, represented as Λ_∞, or Λ_0. Potassium chloride is regarded as a *strong electrolyte*, that is a substance which is completely ionized in solution (and to a large extent in the solid state). The increase in equivalent conductivity with dilution is thought to be due to a lessening of the effects of strong inter-ionic attractive forces, which in higher concentrations prevent the ions from moving freely.

The curve for acetic acid shows a low initial value of equivalent conductivity in concentrated solution, and a very slow increase is seen on dilution, yet the value at infinite dilution is very much higher than that of potassium chloride. This is typical of a *weak electrolyte*, that is a substance which produces very few ions in solution, and which ionizes further as the dilution increases. The ionization of a weak electrolyte can be regarded as an equilibrium process. Taking acetic acid as an example, the number of moles of each constituent at equilibrium is given by:

$$CH_3COOH \rightleftharpoons H^+ + CH_3COO^-$$
$$1-x \qquad x \qquad x$$

in a volume V litres, where the degree of ionization is x. The equilibrium constant for the ionization is, therefore,

$$K_c = \frac{\frac{x}{V} \times \frac{x}{V}}{\frac{1-x}{V}} = \frac{x^2}{V(1-x)}$$

If V increases, then to keep the ratio constant and equal to K, the value of x must increase, that is more ions are produced. At infinite dilution ionization is complete in the case of a weak electrolyte, while for a strong electrolyte ionization is complete at all dilutions, so that infinite dilution signals the point where inter-ionic attractive forces are of no consequence.

Typical strong electrolytes are strong acids (sulphuric, hydrochloric, etc.) strong bases (sodium and potassium hydroxide) and ionic salts (e.g. sodium chloride and sodium acetate). Weak electrolytes include weak acids and bases such as ammonium hydroxide and hydrocyanic acid and most organic acids and bases. For a weak electrolyte AB, if the degree of ionization is x, then at equilibrium, the number of moles of each species present is given by:

$$AB \rightleftharpoons A^+ + B^-$$
$$1-x \quad x \quad x$$

Now the equivalent conductance is proportional to the number of ion-pairs present, i.e. $\Lambda \propto x$; and at infinite dilution, $\Lambda_\infty \propto 1$, since $x = 1$ as the electrolyte is fully ionized at infinite dilution. Hence,

$$\frac{\Lambda}{\Lambda_\infty} = \frac{x}{1} = x$$

Example 67 Find the degree of ionization of acetic acid if an M/5000 solution has an equivalent conductance of 50 ohm^{-1} cm^2 and the value of Λ_∞ is 350 ohm^{-1} cm^2.

Solution For a weak electrolyte,

$$x = \frac{\Lambda}{\Lambda_\infty} = \frac{50}{350} = 0.143$$
$$= 14.3\%$$

(For strong electrolytes which are fully ionized, the ratio Λ/Λ_∞ represents a degree of interference with the free movement of ions through the solution. Sometimes this ratio is referred to as the *apparent degree of ionization*.)

Kohlrausch's law *The equivalent conductivity of an electrolyte at infinite dilution is the sum of the ionic conductivities of the ions produced by that electrolyte.*

The ionic conductivity of the cation is represented by λ_c and that of the anion is λ_a; thus the law is:

$$\Lambda_\infty = \lambda_c + \lambda_a$$

Kohlrausch's law is particularly useful as it allows the equivalent conductivity of a weak electrolyte (at infinite dilution) to be calculated. This is a quantity which is almost impossible to measure practically. For example, λ_{H^+} is 315 ohm^{-1} cm^2 and $\lambda_{CH_3COO^-}$ is 35 ohm^{-1} cm^2. Thus

$$\Lambda_\infty = \lambda_c + \lambda_a = 315 + 35 = 350 \text{ ohm}^{-1} \text{ cm}^2$$

Example 68 The equivalent conductivities of potassium chloride, sodium chloride and potassium nitrate are 130·1, 109·0 and 126·3 ohm^{-1} cm^2 respectively. What is the equivalent conductivity of sodium nitrate?

Solution

$$\Lambda_{KCl} - \Lambda_{NaCl} = 130 \cdot 1 - 109 \cdot 0 = 21 \cdot 1$$

Hence, the ionic conductivity of the sodium ion is $21 \cdot 1$ ohm^{-1} cm^2 less than that of the potassium ion. Therefore,

$$\Lambda_{KNO_3} - \Lambda_{NaNO_3} = 21 \cdot 1$$

or,

$$\Lambda_{NaNO_3} = 126 \cdot 3 - 21 \cdot 1$$
$$= 105 \cdot 2 \text{ ohm}^{-1} \text{ cm}^2$$

The Solubility of Sparingly Soluble Salts

Kohlrausch's law can be used in conjunction with conductivity measurements in the determination of sparingly soluble salts. This is illustrated in Example 69.

Example 69 A saturated solution of silver chloride has a specific conductivity (corrected for the conductivity of water) of $1 \cdot 2 \times 10^{-6}$ ohm^{-1} cm^{-1}. If the ionic conductivities of the chloride ion and silver ion are $65 \cdot 5$ and $54 \cdot 3$ ohm^{-1} cm^2 respectively (at the temperature of the experiment), calculate the solubility of silver chloride.

Solution In these calculations it is assumed that the silver chloride solution is so dilute that conditions of infinite dilution obtain. Using the relation $\kappa V = \Lambda = \Lambda_\infty$ in this case,

$$\Lambda_\infty = \lambda_{Ag^+} + \lambda_{Cl^-}$$
$$= 54 \cdot 3 + 65 \cdot 5$$
$$= 119 \cdot 8 \text{ ohm}^{-1} \text{ cm}^2$$

Hence,

$$V = \frac{119 \cdot 8}{1 \cdot 2 \times 10^{-6}}$$
$$= 9 \cdot 98 \times 10^7 \text{ cm}^3$$

Now V is the volume in cm^3 which contains the equivalent weight of silver chloride, or $143 \cdot 4$ g. Hence, $9 \cdot 98 \times 10^7$ cm^3 or $9 \cdot 98 \times 10^4$ litres contain $143 \cdot 4$ g silver chloride, or one litre contains

$$\frac{143 \cdot 4}{9 \cdot 98 \times 10^4} = 0 \cdot 00144 \text{ g}$$

Thus the solubility of silver chloride is $0 \cdot 00144$ g l^{-1}.

Ostwald's Dilution Law

The variation of conductivity with dilution is explained in terms of the ionic theory as outlined on p. 198. With weak electrolytes, the small extent to which they are ionized produces an equilibrium mixture of ions and un-ionized molecules.

In the case of a 1:1 electrolyte (i.e. one that produces an equal number of cations and anions, such as potassium chloride or magnesium sulphate), the ionization can be represented by:

$$AB \rightleftharpoons A^+ + B^-$$

and

$$K = \frac{[A^+][B^-]}{[AB]}$$

If the degree of ionization is x (i.e. x cations and x anions are produced by the ionization of a fraction x of one mole of electrolyte), then, in a total volume of V litres, the concentration of each species at equilibrium is

$$[AB] = \frac{1-x}{V} \qquad [A^+] = \frac{x}{V} \qquad [B^-] = \frac{x}{V}$$

and

$$K = \frac{\frac{x}{V} \times \frac{x}{V}}{\frac{1-x}{V}} = \frac{x^2}{V(1-x)}$$

If x is very small, then $1-x$ is approximately equal to unity, so that

$$K = \frac{x^2}{V}$$

This relationship is known as *Ostwald's dilution law*; x is the degree of ionization, and K is the dissociation constant.

Example 70 A weak 1:1 electrolyte is 1·8% ionized in M/10 solution. What would the degree of ionization be in an M/100 solution?

Solution

$$K = \frac{x^2}{V} \quad \text{or} \quad \frac{x_1^2}{V_1} = \frac{x_2^2}{V_2}$$

Hence,

$$\frac{(1\cdot 8)^2}{10} = \frac{x_2^2}{100}$$

and
$$x_2 = 10 \times 1\cdot8^2 \times 1\cdot8$$
$$x_2 = 5\cdot7\%$$

Limitations of the Ionic Theory

According to the ionic theory, electrolytes are divided into two classes—strong electrolytes which are regarded as being completely ionized, and weak electrolytes which remain largely un-ionized in the usual concentrations. It is assumed that the ions present in either type of electrolyte behave as point charges (occupying no volume), and suffer no inter-ionic attractions. This is an ideal situation, and the deviations from ideality are most marked in the case of strong electrolytes. At any dilution, the equivalent conductivity of a strong electrolyte corresponds to exactly the same number of ions carrying the current between the two electrodes, since the strong electrolyte is fully ionized. Therefore, the fact that the equivalent conductivity increases with increasing dilution shows that inter-ionic forces of attraction restrict the ions from contributing fully to conductivity (and other chemical functions). Dilution of a strong electrolyte cannot produce any more ions, but it does result in an increase in the distance separating the ions, and this reduces the force of attraction. A quantitative theory of inter-ionic attraction was advanced by Debye and Hückel. However, we are faced with a problem in all cases involving the concentration of ions in solution. The term concentration tells us only how many ions are present per litre of solution—how can we express in a numerical form the extent to which these ions are able to function in a process? The answer is given in terms of an idea expressed in the statement of the law of mass action in which the designation *active mass* is used in this context. Hence, we use the term activity (i.e. active mass in solution) in place of concentration, and the two quantities are related by the equation:

activity = concentration × activity coefficient

In symbols,
$$a = c \times f$$

The activity coefficient f is a factor which varies with the nature of the electrolyte and the dilution of the solution, and it shows the restriction on the complete functioning of an ion due to inter-ionic attractions and other effects. For most strong 1:1 univalent electrolytes, the activity coefficient has values (found by experiment) ranging from 0·8 in 0·1M solution to 0·95 in 0·001M solution. At infinite dilution, f is unity. Consequently, the term concentration when applied to ions or ionic salts (as in this chapter and in Chapter 5) in solution, should be replaced by activity. In an attempt to avoid over-complicating the subject of physical chemistry, activities have not been used in this book, but it should be remembered that the term concentration is an approximation for the quantity we wish to enumerate.

(*Note:* the active mass of a pure solid taking part in a heterogeneous equilibrium is taken as unity and is independent of the quantity of the solid present.)

Acids and Bases

Acids are encountered at an early stage in a chemistry course, and we learn that all acids contain hydrogen atoms which may be replaced by a metal. This gives rise to the definition that an acid is a substance which can produce (hydrated) hydrogen ions in solution. A base is defined as a substance which reacts with an acid to produce a salt and water only, while a soluble base is referred to as an alkali. A broader definition was given by Brønsted and Lowry: *An acid is a species which can donate a proton to a base, while a base is a species which accepts a proton.* This definition may be expressed in the form of an equation:

$$A \rightleftharpoons B + H^+$$
acid   base
(proton donor) (proton acceptor)

A free proton (H^+) does not exist; it is attached by a coordinate link to a lone pair of electrons on the solvent molecule. The proton is then said to be *solvated*, e.g.

$$\overset{\oplus}{H \leftarrow \overset{..}{\underset{..}{O}} : H} \qquad \overset{\oplus \ H}{H \leftarrow \overset{..}{\underset{..}{N}} : H}$$
$$ H \qquad\qquad\qquad H$$

Solvents which readily accept protons (such as water and ammonia) are termed *protophilic*, while solvents which do not accept protons at all (liquid hydrocarbons for example) are termed *aprotic*. The ability of an acid to give up a proton is obviously influenced by the solvent. For instance, water attracts protons readily, and a molecule such as hydrogen chloride functions as a strong acid:

$$HCl + H_2O \rightleftharpoons H_3O^+ + Cl^-$$

In glacial acetic acid, which has a feeble attraction for protons, hydrogen chloride is a weak acid.

$$HCl + CH_3COOH \rightleftharpoons CH_3COOH_2^+ + Cl^-$$

In an aprotic solvent, hydrogen chloride shows no acidic properties; furthermore, in a solvent which is a stronger proton donor than itself, hydrogen chloride could be forced to act as a base. The effect of the solvent on bases is the reverse of the above, basic properties being enhanced by protogenic solvents and so on. All protophilic solvents function as a base, since they can accept protons. Some solvents, such as water and liquid ammonia, are both basic and acidic, undergoing self-ionization as follows:

$$H_2O + H_2O \rightleftharpoons H_3O^+ + OH^-$$

$$NH_3 + NH_3 \rightleftharpoons NH_4^+ + NH_2^-$$

Strengths of Acids

The strength of an acid is measured in terms of its ability to furnish protons to a base. In the same solvent (and we shall discuss aqueous solutions only from now on), the strongest acid is the one which shows the greatest degree of dissociation according to the equation:

$$HA + H_2O \rightleftharpoons H_3O^+ + A^-$$

Note the distinction between a strong acid, defined in terms of degree of dissociation, and a concentrated solution of an acid. For the dissociation (ionization) of one mole of a weak acid in a volume V litres, at equilibrium we have

$$HA \rightleftharpoons H^+ + A^-$$

Moles present at equilibrium $(1-x)$ (x) (x)

(H^+ = hydrated proton)

$$K_a = \frac{x^2}{(1-x)V} = \frac{x^2}{V}$$

if the degree of dissociation, x, is small. Hence

$$x = \sqrt{(K_a V)}$$

Consequently, an indication of the strength of a weak acid can be obtained from the value of its dissociation constant. For instance, the following acids are arranged in order of increasing strengths, as their dissociation constants show.

Phenol $K_a = 1 \times 10^{-10}$ mole l^{-1}
Hydrocyanic acid $K_a = 7 \cdot 2 \times 10^{-10}$ mole l^{-1}
Acetic acid $K_a = 1 \cdot 8 \times 10^{-5}$ mole l^{-1}
Formic acid $K_a = 2 \cdot 1 \times 10^{-4}$ mole l^{-1}

In a similar way, the strengths of weak bases can be compared. For example:

aniline $K_b = 4 \cdot 7 \times 10^{-10}$ mole l^{-1}
ammonia $K_b = 1 \cdot 8 \times 10^{-5}$ mole l^{-1}
methylamine $K_b = 4 \cdot 0 \times 10^{-4}$ mole l^{-1}

The relative strengths of strong acids may be deduced by a variety of methods including:

(a) Conductivities (see p. 200).
(b) The catalytic effect of acids on the hydrolysis of esters. It is found that the rate at which hydrolysis takes place depends on the hydrogen ion concentration. Consequently, a comparison of the rates of hydrolysis of an ester in the presence of acids of equivalent concentrations allows the strengths of the acids to be compared.
(c) Thermochemical measurements. The heat of neutralization of a strong acid by a strong base has a constant value of 56·7 kJ. This is due to the fact that strong acids and bases, and the salts they produce on neutralization, are all fully ionized. Thus, the neutralization reaction is really the formation of water molecules from hydrated hydrogen and hydroxide ions in solution:

$$H^+(\text{hydrated}) + OH^-(\text{hydrated}) \rightarrow H_2O \quad \Delta H = -56 \cdot 7 \text{ kJ}$$

When a weak acid or base is used, the heat change is different from 56·7 kJ.

The Ionic Product of Water

The purest water, conductivity water, is a very feeble conductor of an electric current, and it is concluded that water is ionized to a small extent. This is termed *self-ionization*:

$$H\overset{..}{\underset{..}{O}}{:} \leftarrow H\overset{..}{\underset{..}{O}}{:} \rightleftharpoons \left(H\overset{..}{\underset{H}{O}}{:}H \right)^+ + \left({:}\overset{..}{\underset{..}{O}}{:}H \right)^-$$
$$H H$$

Both the hydroxide and hydrogen ions are hydrated to a greater extent than is indicated by the above equation. Neglecting the hydration of the ions in this equilibrium,

$$H_2O \rightleftharpoons H^+ + OH^-$$

$$K_c = \frac{[H^+][OH^-]}{[H_2O]}$$

Since very few water molecules ionize, the concentration of water, in moles per litre ($1000/18 = 55 \cdot 5$ mole l^{-1}) is regarded as constant. Thus,

$$K_c[H_2O] = K_w = [H^+][OH^-]$$

K_w is known as the ionic product of water, and at 25°C it has a value of 10^{-14} mole^2 l^{-2}. In pure water, the formation of every hydrogen ion is accompanied by the production of a hydroxide ion, so that

$$[H^+][OH^-] = K_w = [H^+]^2$$

Therefore,

$$[H^+] = \sqrt{10^{-14}} = 10^{-7} \text{ mole } l^{-1}$$

The pH Scale

A pH value is of importance in relation to solutions of low hydrogen ion concentration—for example, in solutions of weak acids and bases. pH is defined by the equation

$$pH = -\log_{10}[H^+] = \log_{10}\frac{1}{[H^+]}$$

where the hydrogen ion concentration is expressed in mole l^{-1}.

Example 71 Calculate the pH of M/100 sulphuric acid and M/200 potassium hydroxide if K_w is 10^{-14} mole^2 l^{-2}.

Solution In M/100 sulphuric acid,

$$[H^+] = \frac{2}{100}$$

$$pH = \log\frac{1}{[H^+]} = \log\frac{100}{2} = 1 \cdot 699 = 1 \cdot 7$$

Alternatively,

$$pH = -\log\frac{2}{100} = -(\bar{2} \cdot 3010)$$

$$\begin{aligned}\text{pH} &= -(-2+0{\cdot}3010)\\ &= 2-0{\cdot}3010\\ &= 1{\cdot}7\end{aligned}$$

For M/200 potassium hydroxide,

$$[\text{OH}^-] = \frac{1}{200}$$

Since $[\text{H}^+][\text{OH}^-] = 10^{-14}$, we have

$$[\text{H}^+] = \frac{10^{-14}}{1/200} = 2\times 10^{-12}$$

$$\begin{aligned}\text{pH} = -\log(2\times 10^{-12}) &= -\log 2 - \log 10^{-12}\\ &= -0{\cdot}3010 - (-12)\\ &= 11{\cdot}699 = 11{\cdot}7\end{aligned}$$

The pH of water The concentration of hydrogen ion in pure water at 25°C is 10^{-7} mole 1^{-1}, therefore, the pH of pure water at 25°C is

$$-\log 10^{-7} = 7$$

In pure water, the concentration of hydrogen ion is equal to the concentration of hydroxide ion, and such a condition is termed *neutral*. A solution with a pH lower than 7 is acid, and the lower the pH the more acidic the solution. Similarly, solutions with a pH greater than 7 are alkaline, and the degree of alkalinity increases as the pH increases.

The pH of solutions of weak acids and bases The equilibrium resulting from the ionization of a weak monobasic acid is represented by:

$$\text{HA} \rightleftharpoons \text{H}^+ + \text{A}^-$$

and the dissociation constant is

$$K = \frac{[\text{H}^+][\text{A}^-]}{[\text{HA}]}$$

The dissociation constant is usually designated as K_a to show that it is the dissociation constant of an acid. (Similarly, K_b is used for the dissociation constant of a base.)

In the above ionization:

(a) the concentrations of hydrogen ion and anion are equal: $[\text{H}^+] = [\text{A}^-]$;

(b) very few acid molecules dissociate since the acid is weak, so that the equilibrium concentration of un-ionized acid is not significantly different from the initial concentration (C) of the acid.

Consequently,

$$K_a = \frac{[H^+]^2}{C}$$

or

$$[H^+] = \sqrt{(K_a C)}$$

and

$$-\log [H^+] = -\tfrac{1}{2}\log K_a - \tfrac{1}{2}\log C$$

Hence

$$pH = \tfrac{1}{2}pK_a - \tfrac{1}{2}\log C \quad (\text{where } pK_a = -\log K_a)$$

Example 72 Calculate the pH of 0·1M acetic acid if K_a is $1·8 \times 10^{-5}$ mole l^{-1}.

Solution Using the equation $pH = \tfrac{1}{2} pK_a - \tfrac{1}{2}\log C$,

$$\begin{aligned}
pH &= -\tfrac{1}{2}\log (1·8 \times 10^{-5}) - \tfrac{1}{2}\log 0·1 \\
&= -\tfrac{1}{2} \times 0·255 - \tfrac{1}{2} \times (-5) - \tfrac{1}{2} \times (-1) \\
&= -0·127 + 2·500 + 0·500 \\
&= 2·87
\end{aligned}$$

The pH of a solution of a weak base (BOH) may be deduced in a similar way:

$$BOH \rightleftharpoons B^+ + OH^-$$

$$K_b = \frac{[B^+][OH^-]}{[BOH]}$$

As before, $[B^+] = [OH^-]$, and for a weak base, the equilibrium concentration of the undissociated base is not significantly different from the initial concentration C of the base. Thus

$$\frac{[OH^-]^2}{C} = K_b$$

and

$$[OH^-]^2 = K_b C$$

But

$$[H^+][OH^-] = K_w$$

so that

$$[H^+] = \frac{K_w}{[OH^-]} = \frac{K_w}{\sqrt{(K_b C)}}$$

and
$$-\log[H^+] = -\log K_w + \tfrac{1}{2}\log K_b + \tfrac{1}{2}\log C$$
or
$$pH = 14 + \tfrac{1}{2}\log K_b + \tfrac{1}{2}\log C$$

Example 73 Deduce the pH of a solution of M/100 ammonium hydroxide, given that K_b is $1\cdot7 \times 10^{-5}$ mole l^{-1}.

Solution Using the equation $pH = 14 + \tfrac{1}{2}\log K_b + \tfrac{1}{2}\log C$,

$$\begin{aligned}pH &= 14 + \tfrac{1}{2}\log(1\cdot7\times 10^{-5}) + \tfrac{1}{2}\log 10^{-2}\\ &= 14 + \tfrac{1}{2}\times 0\cdot23 + \tfrac{1}{2}\times(-5) + \tfrac{1}{2}\times(-2)\\ &= 10\cdot61\end{aligned}$$

Alternatively, the pH of a solution can be used to obtain the degree of dissociation or the dissociation constant for a weak electrolyte.

Example 74 If the pH of an M/100 solution of a weak monobasic acid is 5·4, calculate the value of K_a for the acid.

Solution Using the equation $pH = -\tfrac{1}{2}\log K_a - \tfrac{1}{2}\log C$,

$$\begin{aligned}\tfrac{1}{2}\log K_a &= -pH - \tfrac{1}{2}\log C\\ &= -5\cdot4 - \tfrac{1}{2}\log 10^{-2}\\ &= -4\cdot4\end{aligned}$$

and
$$\log K_a = \overline{8}\cdot8000 = \overline{9}\cdot 2000$$
from which,
$$K_a = \text{antilog } \overline{9}\cdot 2000 = 1\cdot6 \times 10^{-9} \text{ mole l}^{-1}$$

Taking the calculation a stage further, if the equilibrium resulting from the ionization of this weak monobasic acid is represented by:

$$\begin{array}{cccc}HA &\rightleftharpoons& H^+ &+ A^-\\ 1-x & & x & x\end{array} \quad \text{moles present in a volume } V$$

then,
$$K_a = \frac{x^2}{V(1-x)} = \frac{x^2}{V}$$

(neglecting $1-x$). Now, the volume of this weak electrolyte solution which contains one mole of HA is 100 litres, so

$$\begin{aligned}x^2 &= 1\cdot6\times 10^{-9} \times 100\\ &= 16\times 10^{-8}\end{aligned}$$

and
$$x = 4 \times 10^{-4} \quad (\text{or } 0.04\%)$$

The Hydrolysis of Salts

Not all salts produce a neutral solution when dissolved in water. This is due to:

(a) the presence of hydrogen and hydroxide ions in water,
(b) the possibility of the formation of a weak acid or a weak base in solution, by reaction of an ion from the salt with either the hydrogen or hydroxide ion from the solvent.

Four different cases arise.

Salt of a strong base and strong acid Sodium chloride is an example. The ions present in solution are:

$$\underbrace{Na^+ + Cl^-}_{\text{from the fully ionized salt}} + \underbrace{H^+ + OH^-}_{\text{from water}}$$

No reaction takes place (the only possible products are HCl and NaOH, both of which are strong electrolytes), so that the concentrations of hydrogen and hydroxide ion remain equal and the solution is neutral.

Salt of a strong base and weak acid Sodium acetate is an example. The ions present in the salt (fully ionized) are Na^+ and CH_3COO^-, and the ions present in pure water are H^+ and OH^-. The reaction

$$CH_3COO^- + H^+ \rightleftharpoons CH_3COOH$$

takes place since the acetic acid formed by this reaction is only feebly ionized. Hence, hydrogen ions are removed from solution, and the excess of hydroxide ion produces an alkaline reaction.

Salt of a weak base and strong acid Ammonium chloride is an example. The ions present in the fully ionized salt are NH_4^+ and Cl^-, and again the ions present in pure water are H^+ and OH^-. The reaction

$$NH^+ + OH^- \rightleftharpoons NH_4OH$$

takes place, since the ammonium hydroxide is a weak electrolyte. This removes hydroxide ions from solution, giving an excess of hydrogen ions and an acidic reaction.

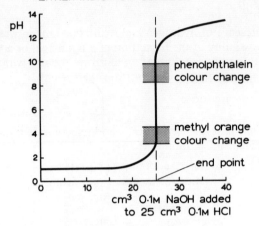

(a) Typical strong acid-strong base titration

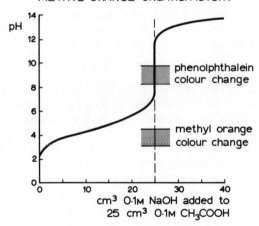

(b) Typical weak acid-strong base titration

Fig. 71

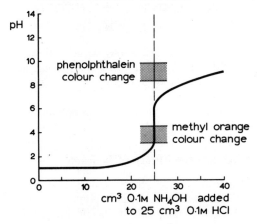

(c) Typical strong acid-weak base titration

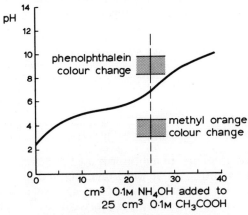

(d) Typical weak acid-weak base titration

Fig. 71

Salt of a weak acid and a weak base Ammonium acetate is an example. The ions present in the salt are NH_4^+ and CH_3COO^-, and as before the ions present in water are H^+ and OH^-. Both hydrogen and hydroxide ions are removed by the reactions

$$H^+ + CH_3COO^- \rightleftharpoons CH_3COOH$$
$$NH_4^+ + OH^- \rightleftharpoons NH_4OH$$

For such salts, the final solution may be neutral, acidic or alkaline, depending on the relative strengths of the weak acids and bases formed by the reactions in solution. In the case of ammonium acetate, the dissociation constants of acetic acid and ammonium hydroxide are very nearly equal, so that a solution of ammonium acetate is neutral, having a pH of 7.

Titration Curves and the Theory of Indicators

An indicator is a substance which changes colour according to the pH of the solution with which it is in contact. Many indicators are weak acids which alter their colour on ionization; for example

$$HX \rightleftharpoons H^+ + X^-$$

initial colour   new colour
of indicator   of indicator

In alkaline solutions, the indicator will be more likely to be present in its ionic form (X^-), while in acid solutions, the un-ionized form predominates. Over a small pH range, a colour change is seen. This is illustrated in the following list.

Indicator	Colour below the pH range	pH range	Colour at a pH above the given range
methyl orange	pink	3·1–4·4	yellow
methyl red	red	4·4–6·2	yellow
cresol red	yellow	7·2–8·8	red
phenolphthalein	colourless	8·3–10·0	pink
thymolphthalein	colourless	10·0–11·6	blue

In acid–alkali titrations, an indicator is used to detect the point (the *end point* of the titration) at which the acid and alkali are in the exact proportions necessary to form salt and water only. Since a solution of a salt in water may not necessarily have a pH of 7, the end point of the titration may not correspond with neutrality (i.e. pH 7). In addition, the pH of the solution must change rapidly as the acid or alkali is added towards the end point of the titration, so that the addition of one drop of titrant is sufficient to cause a distinct change in the colour of the indicator. Hence, the choice of a suitable indicator depends on:

(a) the pH of the salt solution formed as a result of the reaction between the acid and the alkali during the titration,
(b) the rate of change of the pH of the solution near the end point.

The choice of indicator for an acid–alkali titration is made easier by reference to the typical titration curves in Figure 71, which show the pH of the solution as alkali is added to 25 cm^3 of 0·1M monobasic acid.

Buffer Solutions

A buffer solution has two important properties:

(a) It can be prepared to have a definite, calculated pH value.
(b) The pH of such a solution does not change significantly on dilution or on addition of small quantities of an acid or an alkali.

A buffer solution which has a pH less than 7 (an acid buffer) is a solution containing a weak acid and the sodium or potassium salt of that acid; sodium acetate and acetic acid for example. Sodium acetate gives the ions Na$^+$ and CH$_3$COO$^-$, while the acetic acid ionizes slightly:

$$CH_3COOH \rightleftharpoons H^+ + CH_3COO^-$$

Since the acetic acid is very slightly ionized, a large excess of acetate ion (from the sodium acetate) is present. On addition of small amounts of an acid, i.e. hydrogen ion, more un-ionized acetic acid is formed by the reaction between the excess of acetate ion and the added hydrogen ions:

$$\underset{\text{added}}{H^+} + \underset{\substack{\text{present in}\\\text{buffer solution}}}{CH_3COO^-} \rightleftharpoons CH_3COOH$$

Consequently, the pH remains almost unaltered.
    Addition of a small amount of base, i.e. hydroxide ion, brings about the reaction

$$CH_3COOH + OH^- \rightarrow H_2O + CH_3COO^-$$

and again the pH remains virtually unchanged.
    Alkali buffer solutions (having a pH greater than 7) are formed by dissolving the salt of a weak base (e.g. the sulphate, chloride or

nitrate) in a solution of the weak base. A solution of ammonium hydroxide containing ammonium chloride is a typical example. Ammonium chloride gives the ions NH_4^+ and Cl^-, while in solution

$$NH_4OH \rightleftharpoons NH_4^+ + OH^-$$
weakly ionized

The addition of hydroxide ion is counterbalanced by the reaction

$$\underset{\substack{\text{present in}\\\text{excess in}\\\text{buffer solution}}}{NH_4^+} + \underset{\text{added}}{OH^-} \rightleftharpoons NH_4OH$$

while additional hydrogen ions are removed by the reaction

$$NH_4OH + H^+ \rightarrow NH_4^+ + H_2O$$

In both cases, the pH of the solution remains almost unaltered.

The pH of a buffer solution The pH of a buffer solution may be calculated from the value of the dissociation constant of the weak acid or base involved. In the case of the acetic acid–sodium acetate buffer, the sodium acetate consists of ions Na^+ and CH_3COO^-, while the acetic acid is feebly ionized in solution:

$$CH_3COOH \rightleftharpoons CH_3COO^- + H^+$$

The dissociation constant for the acid is

$$K_a = \frac{[H^+][CH_3COO^-]}{[CH_3COOH]}$$

It may be assumed that all the acetate ion concentration stems from the sodium acetate (the salt), so we can say $[CH_3COO^-]$ = [salt]. It may also be assumed that, as the acetic acid is very slightly ionized, the concentration of un-ionized acid is the same as the initial concentration of the acid, i.e. $[CH_3COOH]$ = [acid]. Hence,

$$K_a = \frac{[H^+][\text{salt}]}{[\text{acid}]}$$

or

$$[H^+] = K_a \times \frac{[\text{acid}]}{[\text{salt}]}$$

and

$$\text{pH} = -\log K_a + \log \frac{[\text{salt}]}{[\text{acid}]}$$

(Note the inversion of the ratio [acid]/[salt], which makes the logarithmic term positive.) Example 75 illustrates this type of calculation.

Example 75 Calculate the pH of a solution produced by (a) dissolving 0·1 mole of sodium acetate in 1 litre of M acetic acid, (b) mixing 50 cm³ of 0·1M acetic acid with 150 cm³ of 0·4M sodium acetate solution. (K_a for acetic acid at room temperature is $1·7 \times 10^{-5}$ mole l^{-1}.)

Solution (a)
$$[\text{salt}] = 0·1 \text{ mole } l^{-1}$$
$$[\text{acid}] = 1 \text{ mole } l^{-1}$$

Substituting,
$$\text{pH} = -\log K_a + \log \frac{[\text{salt}]}{[\text{acid}]}$$
$$= -\log 1·7 \times 10^{-5} + \log \frac{0·1}{1}$$
$$= -0·23 - (-5) + \bar{1}·0000$$
$$= 4·77 - 1$$
$$= 3·77$$

(b) On mixing the two solutions, the total volume is 200 cm³. Hence,
$$[\text{salt}] = \frac{0·4 \times 150}{200} \text{ mole } l^{-1}$$
$$[\text{acid}] = \frac{0·1 \times 50}{200} \text{ mole } l^{-1}$$

and
$$\text{pH} = -\log K_a + \log \frac{[\text{salt}]}{[\text{acid}]}$$
$$= -\log 1·7 \times 10^{-5} + \log \left(\frac{0·4 \times 150}{200} \times \frac{200}{0·1 \times 50} \right)$$
$$= 4·77 + \log 12$$
$$= 5·85$$

The pH of an alkaline buffer may be calculated in a similar way. If the weak base is BOH and the salt is BX, then:
$$\text{BOH} \rightleftharpoons \text{B}^+ + \text{OH}^-$$

while BX is B^+ and X^-. Thus

$$K_b = \frac{[B^+][OH^-]}{[BOH]}$$

As before, it is assumed that

$$[B^+] = [salt]$$
$$[BOH] = \text{initial concentration of base}$$
$$= [base]$$

Hence,

$$K_b = \frac{[OH^-][salt]}{[base]}$$

and

$$[OH^-] = K_b \times \frac{[base]}{[salt]}$$

$$[H^+] = \frac{[K_w]}{[OH^-]} = \frac{K_w}{K_b} \times \frac{[salt]}{[base]}$$

and

$$pH = -\log K_w + \log K_b + \log \frac{[base]}{[salt]}$$

$$= 14 + \log K_b + \log \frac{[base]}{[salt]}$$

Example 76 What is the pH of a solution containing 1 mole of ammonium chloride and 0·2 mole of ammonium hydroxide per litre if K_b is $1·8 \times 10^{-5}$ mole l^{-1}?

Solution

$$pH = 14 + \log 1·8 \times 10^{-5} + \log \frac{0·2}{1}$$

$$= 14 + 0·255 - 5 + \bar{1}·301$$
$$= 9·255 - 1 + 0·301$$
$$= 8·556$$

Buffers are used to provide solutions of known pH for checking instruments and indicators, and to control the pH of solutions in which chemical and biochemical reactions are carried out.

In chemical analysis a buffer solution is automatically produced before the metals in Group III of the qualitative analysis scheme are precipitated. The presence of the ammonium hydroxide–ammonium chloride buffer helps to prevent the formation of too high a concentration of hydroxide ion, which would then lead to the precipitation of other metal hydroxides at this stage.

Solubility Product

Substances normally regarded as being insoluble in water do dissolve to a slight extent, as is shown by the fact that the conductivity of such 'solutions' is slightly higher than that of pure water (see Example 54). Silver chloride is an example; some silver and some chloride ions must be present in solution to give the extra conductivity, and the dissolution of this small amount of silver chloride can be regarded as an equilibrium reaction:

$$\underset{\text{undissolved solid}}{(Ag^+ + Cl^-)} \rightleftharpoons \underset{\text{ions forming a saturated solution}}{Ag^+ + Cl^-}$$

For this equilibrium, we can write,

$$K_c = \frac{[Ag^+][Cl^-]}{[Ag^+ Cl^-]}$$

But the active mass of the undissolved salt is taken as constant, say k. Then,

$$K_c = \frac{[Ag^+][Cl^-]}{k}$$

or

$$K_c \times k = K_s = [Ag^+][Cl^-]$$

K_s is known as the *solubility product* for the sparingly soluble salt. Since the value of an equilibrium constant varies with temperature, so the value of a solubility product alters with temperature.

Calcium fluoride is another example of a sparingly soluble salt. Its solubility product is given by

$$K_s = [Ca^{2+}][F^-]^2$$

The solubility product principle applies only to sparingly soluble salts, and the solubility product is an indication of the maximum concentrations of the ions constituting the salt that can remain in solution. For example, the solubility product of silver chloride is 10^{-10} mole^2 l^{-2}; in a solution where the silver ion concentration is 10^{-10} mole per litre, and the chloride ion concentration is 10^{-2} mole per litre, precipitation will not take place, as the product of these concentrations does not exceed the solubility product:

$$[Ag^+][Cl^-] = 10^{-10} \times 10^{-2} = 10^{-12} \text{ mole}^2 \text{ l}^{-2}$$

Solubility and Solubility Product

The solubility product of zinc sulphide is 10^{-22} mole^2 l^{-2} at a certain temperature. That is,
$$[Zn^{2+}][S^{2-}] = 10^{-22} \text{ mole}^2 \text{ l}^{-2}$$
Provided all the zinc and sulphide ions originate from the dissolution of zinc sulphide, and that they are not brought into solution from any other source, then
$$[Zn^{2+}] = [S^{2-}]$$
so that
$$[Zn^{2+}][S^{2-}] = [Zn^{2+}]^2 = [S^{2-}]^2 = 10^{-22} \text{ mole}^2 \text{ l}^{-2}$$
and
$$[Zn^{2+}] = [S^{2-}] = \sqrt{(10^{-22})} = 10^{-11} \text{ mole l}^{-1}$$

Now the value of either the zinc ion concentration $[Zn^{2+}]$, or the sulphide ion concentration $[S^{2-}]$, is numerically equal to the number of moles of zinc sulphide that have dissolved, so that the solubility of zinc sulphide is
$$s = \sqrt{K_s} = 10^{-11} \text{ mole l}^{-1}$$

Example 77 Calculate the solubility of silver acetate in moles per litre if the solubility product at 30°C is 4.9×10^{-3} mole^2 l^{-2}.

Solution
$$\underset{\text{undissolved}}{CH_3COO^-Ag^+} \rightleftharpoons \underset{\text{ions in solution}}{Ag^+ + CH_3COO^-}$$

$$K_s = [Ag^+][CH_3COO^-] = 4.9 \times 10^{-3} = 49 \times 10^{-4} \text{ mole}^2 \text{ l}^{-2}$$

Since the acetate and silver ions are produced in equal quantity in solution when the salt dissolves, the solubility of silver acetate is
$$s = [Ag^+] = [CH_3COO^-] = \sqrt{K_s} = \sqrt{(49 \times 10^{-4})}$$
$$= 7 \times 10^{-2} \text{ mole l}^{-1}$$

Common Ion Effect

If quantities of a substance which has an ion in common with either the anion or the cation constituting the sparingly soluble salt are added to a solution of the salt, the solubility of the sparingly soluble salt is reduced. This behaviour is known as *the common ion effect*.

Example 78 What is the solubility of silver chloride in (a) pure water, (b) 0·1M silver nitrate solution, (c) M/1000 hydrochloric acid if the solubility product of silver chloride at room temperature is 10^{-10} mole^2 l^{-2}?

Solution (a) The solubility in water is given by
$$s = \sqrt{K_s} = \sqrt{10^{-10}} = 10^{-5} \text{ mole l}^{-1}$$

(b) In 0·1M silver nitrate, the concentration of silver ion from the silver nitrate is 10^{-1} mole l^{-1}. Since $K_s = 10^{-10} = [Ag^+][Cl^-]$, we have
$$[Cl^-] \times 10^{-1} = 10^{-10}$$
$$[Cl^-] = 10^{-9} \text{ mole l}^{-1}$$

The chloride ion concentration comes solely from the dissolved silver chloride, so that, in this case, where the common ion is the cation, the chloride ion concentration is numerically equal to the solubility of silver chloride. Thus the solubility of silver chloride in 0·1M silver nitrate is 10^{-9} mole l^{-1}.

(c) In M/1000 hydrochloric acid, $[Cl^-] = 10^{-3}$, so that
$$[Ag^+] \times 10^{-3} = 10^{-10}$$
and
$$[Ag^+] = 10^{-7} \text{ mole l}^{-1}$$

Since the only source of silver ion in solution is the dissolution of silver chloride, the solubility of silver chloride in M/1000 hydrochloric acid is 10^{-7} mole l^{-1}.

The addition of a solution containing a common ion—the chloride ion in case (c) and the silver ion in case (b)—reduces the solubility of a sparingly soluble salt, from 10^{-5} to 10^{-7} or 10^{-9} moles per litre in the above example involving silver chloride. Other examples of this effect are the formation of a precipitate of barium chloride when concentrated hydrochloric acid is added to a concentrated solution of barium chloride, and the use of solutions containing a common ion for washing precipitates in gravimetric analysis.

Solubility Product and Qualitative Analysis

The precipitating agent for the metals in Groups II and IV of the qualitative analysis scheme is sulphide ion. This is produced by the ionization of hydrogen sulphide:
$$H_2S \rightleftharpoons H^+ + HS^- \rightleftharpoons 2H^+ + S^{2-}$$

In Group II, the solution already contains a high concentration of hydrogen ion, the effect of which is to suppress the ionization of the dissolved hydrogen sulphide. Consequently, only those metal sulphides which have very low solubility products are precipitated, for example:

$$HgS \quad K_s = 3 \times 10^{-54} \text{ mole}^2 \text{ l}^{-2}$$
$$CuS \quad K_s = 3 \times 10^{-42} \text{ mole}^2 \text{ l}^{-2}$$
$$CdS \quad K_s = 3 \times 10^{-29} \text{ mole}^2 \text{ l}^{-2}$$
$$PbS \quad K_s = 4 \times 10^{-28} \text{ mole}^2 \text{ l}^{-2}$$

In Group IV, the solution is alkaline, and ionization of the hydrogen sulphide is enhanced by the removal of hydrogen ions. Therefore, metal sulphides having a higher solubility product are precipitated at this stage, for example:

$$ZnS \quad K_s = 5 \times 10^{-24} \text{ mole}^2 \text{ l}^{-2}$$
$$MnS \quad K_s = 1 \cdot 8 \times 10^{-15} \text{ mole}^2 \text{ l}^{-2}$$
$$CoS \quad K_s = 3 \times 10^{-26} \text{ mole}^2 \text{ l}^{-2}$$
$$NiS \quad K_s = 4 \times 10^{-21} \text{ mole}^2 \text{ l}^{-2}$$

EXERCISES

1. An electric current is passed through two voltameters connected in series, containing respectively, a solution of dilute sulphuric acid (platinum electrodes) and a copper sulphate solution (copper electrodes). What weight of copper will be deposited in the second voltameter if 110 cm³ of hydrogen, measured dry at 15°C and 737 mm pressure, was produced in the first voltameter?
2. The resistance of an M/50 solution of potassium chloride was found to be 250 ohm using conductivity cell A and 295 ohm using another cell B, both at 25°C. If the specific conductivity of M/50 potassium chloride at this temperature is $2 \cdot 78 \times 10^{-3}$ ohm^{-1} cm^{-1}, find the cell constants of A and B, and calculate the equivalent conductivity of M/50 potassium chloride.
3. 500 cm³ of an M/10 solution of sodium chloride was electrolysed by passing a steady current of 0·2 A through the solution for 1 h 36·5 min. Calculate, (a) the weight of chlorine liberated, (b) the weight of sodium hydroxide produced in the solution, (c) the volume of hydrogen liberated at s.t.p.

    Find also, the weight of chlorine obtained by passing a current of 2 A through 500 cm³ of M/10 sodium chloride solution for one hour.
4. The cell constant of a certain conductivity cell is 1·08, and, using this cell, the resistance of a solution of sulphuric acid containing 14 g of H_2SO_4 per litre, is found to be 18·9 ohm. What is the equivalent conductivity of sulphuric acid at this dilution?

5. The equivalent conductivity of sodium nitrate at various dilutions is as follows:

Dilution (litres)	500	1000	2000	3000	4000	5000	∞
Equivalent conductivity	103·1	103·5	103·9	104·1	104·2	104·3	105·8

Plot these figures on a graph, and deduce the apparent degree of ionization of sodium nitrate at a dilution of 750 litres.

6. A saturated solution of barium sulphate has a specific conductivity of $3·73 \times 10^{-6}$ ohm^{-1} cm^{-1} at room temperature. The specific conductivity of pure water is $1·23 \times 10^{-6}$ ohm^{-1} cm^{-1}, and the equivalent conductivity of barium sulphate is 125 ohm^{-1} cm^{2} at this temperature. Calculate the solubility of barium sulphate in grammes per litre at this temperature.

7. The equivalent conductivities at infinite dilution of hydrochloric acid, sodium chloride and sodium acetate are 370, 108·4 and 85·7 ohm^{-1} cm^{2} respectively. What is the equivalent conductivity of acetic acid at infinite dilution? If the specific conductivity of M/10 acetic acid solution is $1·13 \times 10^{-6}$ ohm^{-1} cm^{-1}, what is the degree of ionization of M/10 acetic acid?

8. If the degree of ionization of M/10 ammonium hydroxide is 0·013%, what is the degree of ionization of M/200 solution?

9. The specific conductivity (corrected for the conductivity of water) of silver chloride is $1·13 \times 10^{-6}$ ohm^{-1} cm^{-1}. If the ionic conductivities of silver and chloride ions are 54·0 and 65·2 ohm^{-1} cm^{2} respectively, deduce the solubility of silver chloride in grammes AgCl per litre.

10. Calculate the pH of (a) M/20 potassium hydroxide, (b) a solution containing 9·8 g of sulphuric acid per litre. Assume the apparent degree of ionization is 100% in each case.

11. The equivalent conductivity of M/1000 acetic acid is 53·8 ohm^{-1} cm^{2} and the equivalent conductivity of acetic acid at infinite dilution is 384 ohm^{-1} cm^{2}. Calculate the degree of ionization of M/1000 acetic acid, and by using Ostwald's dilution law, deduce the degree of ionization at a dilution of 10000 litres.

12. Calculate the pH of an M/10 solution of ammonium hydroxide if K_b is $1·73 \times 10^{-5}$ mole l^{-1}.

13. The pH of a solution is 4·4. What is the hydrogen ion concentration of this solution?

14. The pH of an M/10 solution of a weak monobasic acid is 3·25. What is its dissociation constant?

15. Calculate the pH of a solution formed by mixing 10 cm^{3} of M/10 acetic acid with 90 cm^{3} of M/10 sodium acetate solution. (K_a for acetic acid is $1·7 \times 10^{-5}$ mole l^{-1}.)

16. The specific conductivity of silver bromide is 6.2×10^{-8} ohm^{-1} cm^{-1} (corrected for the conductivity of water) at 20°C. The equivalent conductivity of silver bromide solution at infinite dilution is 124 ohm^{-1} cm^2. Deduce the solubility product of silver bromide at 20°C.
17. If the concentration of hydroxide ion in an M/100 solution of a weak monobasic acid is 10^{-10} mole l^{-1}, calculate the dissociation constant of the acid.
18. The solubility product of cadmium sulphide is 10^{-30} mole^2 l^{-2}. Calculate the solubility of cadmium sulphide (in mole l^{-1}) in (a) water, (b) a solution containing 10^{-5} mole l^{-1} of sulphide ion.
19. How many moles of sodium acetate must be dissolved in one litre of M/20 acetic acid to give a buffer solution of pH 4? (K_a for acetic acid is 1.8×10^{-5} mole l^{-1}.)
20. The solubility product of lead sulphate at room temperature is 1.6×10^{-8} mole^2 l^{-2}. What weight of lead sulphate dissolves when 0.2 g of lead sulphate precipitate is washed with 250 cm^3 portions of (a) water, (b) M/10 sulphuric acid?
21. Given

$$Fe^{3+} + e^- = Fe^{2+} \qquad E^0 = 0.76 \text{ V}$$
$$Cl_2 + 2e^- = 2Cl^- \qquad E^0 = 1.36 \text{ V}$$
$$ClO_4^- + 2H^+ + 2e^- = ClO_3^- + H_2O \qquad E^0 = 1.2 \text{ V}$$

determine whether chlorine is capable of oxidizing (a) Fe^{2+}, (b) ClO_3^-. Also, calculate the E^0 value for the reaction between ClO_4^- and Fe^{2+}.
22. If the standard free energy change for the half-reaction $Cu^{2+} + 2e^- = Cu$ is $\Delta G^0 = -67.5$ kJ mole^{-1}, calculate the corresponding E^0 value.
23. Deduce the E value of the half-reaction $Zn^{2+} + 2e^- = Zn$ if $E^0 = -0.76$ V, at 25°C if the concentration of the zinc ions in solution is 10^{-4}M.

7. The Colloidal State

Particle Size and Colloidal Solutions

The size of the particles is the factor which differentiates between true solutions, colloidal solutions and suspensions.

Suspensions contain solute particles which are large enough to be visible to the naked eye, or under a microscope, and they may be removed by filtration or by centrifuging. A suspension appears milky.

True solutions contain solute particles which are of the same order of size as the solvent particles. The solute particles cannot be discerned under the microscope, they pass through the pores of a filter, and they are not removed by centrifuging. A true solution appears perfectly clear, even when rays of light are made to converge within it.

Colloidal solutions (or *sols*) contain particles which are so small that they are invisible to the eye, even under a microscope, and they pass through the pores of a filter. A colloidal solution appears clear (or very nearly so), but the path of a converging beam of light is clearly seen as a kind of milkiness within the solution—i.e. although the particles themselves cannot be seen, the effects of their presence can be made visible. It was discovered by Graham, in 1861, that a true solution will pass unchanged through a parchment membrane, whereas with a colloidal solution only the solvent will diffuse through the membrane.

The sizes of the particles in true and colloidal solutions vary from about 10^{-7} to 10^{-5} cm (see Figure 72). The total surface area of a given mass of solid increases as it is divided more and more finely. For example, each face of a cube of side 2 cm has an area of 4 cm^2, making a total surface area of 24 cm^2 for the six faces. If this cube is subdivided to make eight cubes, each of side 1 cm, the surface area of each face of the small cube is 1 cm^2, or 6 cm^2 for all the faces of one cube. But there are now eight of these small cubes, so that the total surface area is 48 cm^2. If this subdivision is continued to give cubes of side 10^{-5} cm the total surface area of the original mass of solid is increased to about 5 million cm^2. The effect of this large surface area is of particular importance in connection with the electrical properties of colloids.

For colloidal solutions, the terms solvent and solute are replaced

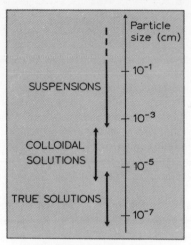

Fig. 72. Range of particle sizes

by the names *dispersion medium* and *disperse phase* respectively, thus implying that the system is heterogeneous. The dispersion medium and the disperse phase may be present as a solid, liquid or a gas, as shown in the following examples.

Dispersion medium	Disperse phase	Type
gas	solid	smoke
gas	liquid	mist, fog
liquid	solid	sols, gels
liquid	liquid	emulsion
liquid	gas	foam
solid	solid	solid sol, solid gel, glass
solid	liquid	solid emulsion
solid	gas	solid foam

The colloidal system most frequently met with is that in which the dispersion medium is a liquid and the disperse phase is a solid, and this system is usually divided into two classes:

1. *Lyophobic* (or solvent-hating) sols, in which there is little interaction between the dispersion medium and the disperse phase. Once the particles held in suspension are coagulated or precipitated, the reverse process (peptization) cannot be brought about simply by stirring or shaking with the dispersion medium, and

for this reason, lyophobic sols are often termed *irreversible sols*. Typical examples are gold, iron (III) hydroxide and sulphur sols.
2. *Lyophilic* (or solvent-liking) sols, in which there is an affinity between the disperse phase and the dispersion medium. The viscosity of such sols is usually higher than the pure dispersion medium, and they may set to a gel. If a gel liquefies on stirring or shaking, the system is said to be *thixotropic*. Lyophilic sols are stable and not easily coagulated, but the sol easily reforms when the coagulum is remixed with the pure dispersion medium. Lyophilic sols are often called *reversible sols*. Typical lyophilic sols are gelatin and starch.

Properties of Colloids

Two general features of colloidal solutions are especially significant: their optical and electrical properties.

Optical properties A beam of light when passed through a colloidal solution is rendered visible (the *Tyndall effect*) because the light is

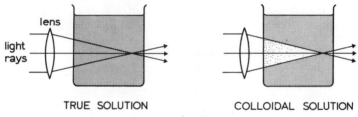

Fig. 73. Tyndall cone

scattered by the colloidal particles. This is similar to the scattering that delineates the beam from a car headlamp on a misty night or the rays from the projector to the screen in the smoke haze of a cinema. When the light rays are made, by means of a lens, to converge within a colloidal solution (Figure 73), the resulting cone-shaped outline is known as the *Tyndall cone*, and this effect cannot be produced in a true solution. If the scattered light is viewed from above by means of a microscope, the points of light resulting from the scattering by the individual particles can be seen. This forms the basis of *Zsigmondy's ultra-microscope*. The ultra-microscope shows that the colloidal particles are in continuous, random movement known as *Brownian motion*.

Electrical properties When a colloidal solution is placed in a

U-tube fitted with electrodes as shown in Figure 74, the colloidal particles move towards one or other of the electrodes, showing that the colloidal particles carry an electrical charge. The movement of colloidal particles under the influence of an electric potential is called *electrophoresis* or *cataphoresis*. Iron (III) hydroxide and aluminium hydroxide colloidal solutions carry a positive charge, while colloidal solutions of metals and arsenic (III) sulphide carry a negative charge. Since a colloidal solution is as a whole electrically neutral, the colloidal particles must have picked up some ions (either positive or negative) preferentially from the dispersion medium. These adsorbed ions are retained quite strongly on the surface of the colloidal particles, and it follows that the dispersion medium must contain an excess of ions of the opposite charge. Therefore, during electrophoresis, the colloidal particles move towards one electrode

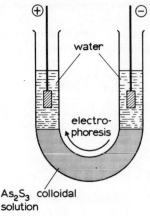

Fig. 74. Electrophoresis

and the dispersion medium moves towards the other. Consequently, if the colloidal particles were to be prevented from moving, the dispersion medium would continue to move towards an electrode and build up a hydrostatic head in so doing. This effect, known as *electro-osmosis*, can be demonstrated.

The Stability of Colloidal Solutions

Because of the adsorption of ions onto the surface of a colloidal particle, each colloidal particle carries a charge. This in turn attracts a layer of ions of opposite charge close to the particle, building up what is known as a *Helmholtz double layer* (Figure 75).

The stability of a colloidal solution is due to the repulsive forces set up between the charged particles, and it is easy to see why the stability of a colloidal solution is enhanced by the presence of a small amount of electrolyte in the dispersion medium. Addition of further quantities of electrolyte neutralizes the charge on the colloidal particles; the point at which this happens is called the *iso-electric point*. At this point, a colloidal solution tends to precipitate.

Neutralization may be brought about:

(a) by allowing a colloidal solution to come into contact with an electrode carrying a charge opposite to that on the colloidal particles;
(b) by mixing two oppositely charged colloidal solutions;
(c) by adding an excess of an electrolyte. In this method, it is found that the coagulating power of an electrolyte depends on the size of the charge carried by the ions—the highly charged ions such as Al^{3+} being the most effective. (This is the essence of the Hardy–Schulze rule.) Hence, aluminium salts are used in preparations to stop bleeding from small cuts by coagulation, as blood is a negative colloid.

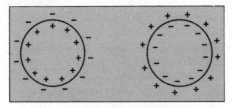

Fig. 75. Helmholtz double layer

Lyophilic colloids are much more resistant to precipitation, and a lyophilic colloid may be added to a lyophobic colloid (of the same charge) to stabilize the latter. This is termed *protective action*, and it can be measured in terms of a gold number, which is the quantity of protective colloid which must be added to a standard gold colloid in order to prevent its coagulation by a standard quantity of sodium chloride.

The Preparation of Colloidal Solutions

The production of colloidal particles—which are intermediate in size between particles in suspension and particles in true solution—may be effected by:

(a) reduction methods, in which large particles are broken down to colloidal dimensions;
(b) aggregation methods, whereby particles in a true solution are coagulated to build up to the colloidal size.

Reduction methods One method is by means of the *colloid mill*: the finely ground material is mixed with the dispersion medium and subjected to further milling between steel rollers until the colloidal size is reached.

The most important reduction method is *Bredig's Arc Method*; in particular it is used in the preparation of colloidal solutions of metals. An arc is struck between two metal electrodes held under the surface of the dispersion medium, in the presence of a small quantity of electrolyte.

Precipitates may be brought into colloidal solution by *peptization*, i.e., by adding an electrolyte which provides ions which may be adsorbed by the precipitate particles. This leads to a breakdown of the precipitate particles due to the repulsive effects of the charges. For example, addition of potassium chloride or silver nitrate to a silver chloride suspension produces a colloidal solution of silver chloride, while the passage of hydrogen sulphide through a cadmium sulphide suspension results in the formation of a cadmium sulphide sol.

Aggregation methods These methods are more commonly used than techniques involving a reduction in particle size. They depend on the fact that when an insoluble substance is produced rapidly by a chemical reaction in a dilute solution, the particles formed tend to be in the colloidal size range. Some typical examples are as follows.

(a) A sulphur sol is produced by the acidification of dilute sodium thiosulphate solution:

$$Na_2S_2O_3 + 2HCl = 2NaCl + SO_2 + H_2O + S$$

(b) A colloidal solution of iron (III) hydroxide is formed when a small quantity of iron (III) chloride is stirred into boiling water.
(c) A gold (or silver) sol may be produced by reducing a dilute solution of the metal salt with a suitable reducing agent (usually an organic reducing agent) such as formaldehyde or a tartrate.
(d) Silica gel is produced by the acidification of a sodium silicate solution.
(e) Passing hydrogen sulphide through a very dilute solution of

arsenic (III) oxide produces a yellow arsenic (III) sulphide sol.
(f) When a solution of sulphur in alcohol is added to water, the liberated sulphur forms a colloidal solution.

Lyophilic sols such as those of starch or gelatin may be prepared by mixing with hot water. A gel may form on cooling, but a sol is re-formed on heating.

The ultra-centrifuge The rate at which particles fall out of suspension depends on the force of gravity. Settling of the particles is accelerated if the gravitational force acting on the particle is augmented by inertial forces, and this can be done very effectively by means of a centrifuge. In order to bring about the sedimentation of colloidal particles, a very high speed centrifuge, the ultra-centrifuge, is used; it is capable of producing a settling force up to 500 000 times that of gravity.

8. Radioactivity

Radioactive substances emit radiations which show the following properties:

- (a) They affect a photographic emulsion.
- (b) They pass through opaque material.
- (c) They are unaffected by temperature and pressure.
- (d) They produce scintillations on a sensitive screen.
- (e) They cause ionization of gases (for instance, they produce visible tracks in the saturated vapour of a cloud chamber).

The Three Types of Radiation

α-rays These rays are deflected by electric and magnetic fields, and it is thus possible to measure the charge-to-mass ratio of the particles of which they consist. The α-particle has been shown to consist of two protons and two neutrons—that is a helium nucleus, He^{2+}. These particles have a low penetrating power and are stopped by a thin metal foil or a piece of card as thick as a postcard.

β-rays These rays are deflected in the opposite sense to α-particles by electric and magnetic fields, and they have been identified as high-speed electrons. They are more penetrating than α-rays, as they will pass through thin metal foil. Lead foil, or plastic sheet (at least $\frac{1}{4}$ inch thick) stops the passage of these rays.

λ-rays This type of radiation is not deflected by magnetic or electric fields. Thus, γ-rays carry no charge, and they have been identified as electromagnetic radiation of shorter wavelength (therefore of higher energy and much more dangerous) than X-rays. They have great penetrating power and the activity due to this type of radiation cannot be completely stopped. Thick lead shielding can reduce the danger to minimal proportions.

Radioactive Decay

The radiations emitted by a radioactive substance may be detected by a Geiger-Müller counter or by a scintillation counter. The Geiger-Müller counter acts as follows. When the radiation passes through the gas kept at low pressure in the Geiger-Müller tube the gas is ionized and allows the momentary passage of an electrical current between two electrodes placed in the gas. This pulse of current is recorded by a suitable electronic device as a *count* which is registered on an instrument known as a *scaler*. The pulse may also be converted into an audible 'click' emitted by a loudspeaker. In a scintillation counter, the count corresponds to a flash of light produced when radiation impinges on a screen coated with a suitable phosphorescent material. A photo-electric cell can be used to convert the flash of light into an electric signal proportional to the intensity of radiation; this signal is fed to a recording instrument. In both methods, the level of radioactivity is expressed in terms of disintegrations per second (d.p.s.) or per minute, or as counts per second (c.p.s.) or per minute.

A graph showing the variation of the radioactivity of a given sample with time shows an exponential decrease in the level of the activity (Figure 76). This curve is known as a *radioactive decay*

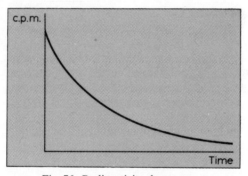

Fig. 76. Radioactivity decay curve

curve, and it is clear from this graph that the time required for the complete decay of a radioactive sample is infinite, as the curve approaches the time axis asymptotically. In theory, the radioactivity never quite disappears, but it reaches such a low level as to be undetectable. The rate of decay is characterized by the *half life period*; this is the time required for the radioactivity of a given sample to decay to half the initial level. Each radioactive species decays at a different rate, and a wide range of half life times ranging from fractions of a second to many millions of years are known. For example, the half life of radon is almost four days. Thus, in four days, half the original number of radon atoms will have disintegrated while, after eight days, the activity will have fallen to one-quarter of the initial value, as shown in Figure 77.

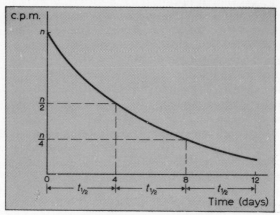

Fig. 77. Decay curve for radon

If n is the number of counts per minute (which is proportional to the number of atoms not yet disintegrated), then after the half life period ($t_{\frac{1}{2}}$) has elapsed, the count rate will be $n/2$. From this, it follows that any point in time will serve as the initial value. (This is an over-simplified picture for the case of radon, as the product of the disintegration of radon is also radioactive, giving rise to a more complicated graph than that shown in Figure 77.)

The Mathematics of Radioactive Decay

The rate of decay (i.e. the activity) is the number of atoms disintegrating per unit time or dN/dt, where N is the number of undecayed atoms at any instant. Now, the rate of decay at any instant is proportional to the number of undecayed atoms present at that instant:

$$-\frac{dN}{dt} \propto N \quad \text{or} \quad -\frac{dN}{dt} = \lambda N$$

λ, the constant of proportionality, is known as the radioactive decay constant, while the negative sign in front of the differential shows that the level of activity decreases as time proceeds.

Rearranging,

$$\frac{dN}{N} = -\lambda\, dt$$

and integrating,

$$\log_e N = -\lambda t + c$$

where c is the constant of integration. When t is zero, the original number of atoms N_0 are present, so

$$\log_e N_0 = c$$

The complete equation is thus,

$$\log_e N_0 - \log_e N = \lambda t$$

or, using logarithms to base 10,

$$\log N_0 - \log N = \frac{\lambda t}{2 \cdot 303}$$

The half life period $t_{\frac{1}{2}}$ corresponds to the instant when just half of the original atoms have decayed, i.e. $N = \frac{1}{2} N_0$. Thus

$$\log N_0 - \log \tfrac{1}{2} N_0 = \frac{\lambda t_{\frac{1}{2}}}{2 \cdot 303}$$

$$\log 2 = 0 \cdot 3010 = \frac{\lambda t_{\frac{1}{2}}}{2 \cdot 303}$$

or

$$t_{\frac{1}{2}} = \frac{0 \cdot 693}{\lambda}$$

Example 79 What is the value of the radioactive decay constant for carbon-14 if the half life of this isotope is 5600 years?

Solution If $t_{\frac{1}{2}} = 0 \cdot 693/\lambda$, then

$$\lambda = \frac{0 \cdot 693}{t_{\frac{1}{2}}} = \frac{0 \cdot 693}{5600} = 1 \cdot 237 \times 10^{-4} \text{ years}^{-1}$$

The curie This is a unit of radioactivity (formerly defined as the activity of one gramme of radium) now standardized as $3 \cdot 7 \times 10^{10}$ d.p.s. A millicurie (one thousandth of a curie) and a microcurie (one millionth of a curie) are also commonly used units.

Example 80 Using the data given in Example 79, calculate the weight of one millicurie of carbon-14. Take Avogadro's number as 6×10^{23}.

Solution The radioactivity of a sample is dN/dt. In this case, one millicurie of radiation is

$$3 \cdot 7 \times 10^7 \text{ d.p.s.} = 3 \cdot 7 \times 10^7 \times 365 \times 24 \times 60 \times 60 \text{ disintegrations per year}$$

Since $dN/dt = \lambda N$ (neglecting the minus sign) N, the number of atoms constituting one millicurie, is

$$\frac{3\cdot 7 \times 10^7 \times 365 \times 24 \times 60 \times 60}{1\cdot 237 \times 10^{-4}} = 9\cdot 43 \times 10^{18}$$

Now 6×10^{23} atoms weigh 14 g, so, $9\cdot 43 \times 10^{18}$ atoms weigh

$$\frac{14 \times 9\cdot 43 \times 10^{18}}{6 \times 10^{23}} = 2\cdot 2 \times 10^{-4} \text{ g}$$

Hence $2\cdot 2 \times 10^{-4}$ g of carbon-14 represents one millicurie.

The Group Displacement Law

The more common disintegration processes involve the ejection of an α- or a β-particle *directly from the nucleus*.

Loss of an α-particle Since an α-particle is composed of two neutrons and two protons, the resulting nucleus is *four units less in mass* and has an *atomic number two less* than the original. For example,

$$^{238}_{92}\text{U} \rightarrow \,^{4}_{2}\text{He}(\alpha) + \,^{234}_{90}\text{Th} \tag{1}$$

or in general,

$$^{A}_{Z}\text{X} \rightarrow \,^{4}_{2}\text{He}(\alpha) + \,^{A-4}_{Z-2}\text{Y}$$

where A is the atomic weight and Z the atomic number of the original element, and Y is the resulting (or daughter) element.

Loss of a β-particle The emission of a β-particle by a nucleus is thought to be the result of a sequence of changes during which a neutron is converted into a proton:

$$\text{n} \rightarrow \text{p}^+ + \beta^- \text{ (electron)}.$$

Thus, no heavy particles are lost by the nucleus and the atomic weight is unaltered. However, the atomic number, which is equal to the number of protons in the nucleus, is increased by one. For example, the thorium nucleus produced as in equation (1) above, itself decays by loss of a β-particle:

$$^{234}_{90}\text{Th} \rightarrow \beta + \,^{234}_{91}\text{Pa} \tag{2}$$

In general, as a result of the loss of a β-particle from the nucleus, the daughter element has the same mass, but is situated one place to the right of the original element in the Periodic Table:

$$^{A}_{Z}\text{X} \rightarrow \beta + \,^{A}_{Z+1}\text{Y}$$

The emission of γ-radiation does not affect either the atomic weight or the atomic number of the element.

The nucleus $^{234}_{91}\text{Pa}$, produced by decay of thorium as in equation (2), is also radioactive, losing another β-particle:

$$^{234}_{91}\text{Pa} \rightarrow \beta + ^{234}_{92}\text{U}$$

As a result of this disintegration, an atom of uranium has been produced. This atom is four units in mass lighter than the original uranium atom of equation (1). The difference between these atoms lies in the fact that two neutrons were lost during the α-particle emission, and two more neutrons were converted into two protons during the subsequent emission of two β-particles. Hence, ^{234}U has four neutrons less in its nucleus than ^{238}U. Atoms of the same element, such as these, which differ only in atomic weight (as a consequence of a differing neutron content in the nucleus) are called *isotopes*. The effects of α- and β-particle emission on the atomic weight and the position of an element in the Periodic Table are summed up by the *group displacement law*:

(a) Following α-particle emission, the daughter element is situated two places to the left of the parent element in the Periodic Table, and is four units lighter in atomic mass.
(b) Following β-particle emission, the daughter element is situated one place to the right of the parent in the Periodic Table, and no change in the atomic weight takes place.

Elements of atomic number greater than 83 (i.e. following bismuth in the Periodic Table) are naturally radioactive—that is, they have no stable isotope. Nuclei of the lighter elements can be made unstable by bombardment with particles such as neutrons or protons (the nucleus absorbs one or more of the bombarding particles), giving rise to induced or artificial radioactivity.

The successive disintegrations of a naturally radioactive element form a *disintegration series*; the first part of the series starting with uranium-238 is given in Figure 78. From radon, the series continues by way of the emission of a total of four α- and four β-particles. Thus, the final stable isotope has an atomic weight of $222 - (4 \times 4)$ or 206, and an atomic number of $86 - (4 \times 2) + (4 \times 1)$ or 82. The end product is therefore an isotope of lead, atomic weight 206.

The Mass Spectrograph

This apparatus can be used for the measurement of isotopic masses. In the diagram of the instrument shown in Figure 79, the rays

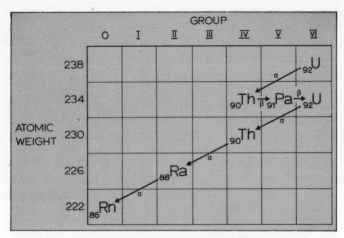

Fig. 78. Natural disintegration series

travelling from cathode to anode in the gas discharge tube are cathode rays. If a perforated cathode is used, an additional set of rays—positive rays—travelling in the opposite direction to the cathode rays, is produced. The nature of these rays depends on the gas in the tube. For instance, using neon, the positive rays consist of neon ions Ne^+ (and a few Ne^{2+}). After passing through the focussing slits S, the narrow beam of positive rays move between the electrodes E, which give rise to an intense electric field. The beam of positive rays is deflected by this field, the lighter particles being turned further from the original path. If the beam consists of two isotopes, say neon-20 and neon-22, the positive ray stream is split into two, shown as H (heavy—neon 22) and L (light—neon 20) in Figure 79. However, the positive particles in the original beam may not all have the same velocities, and while passing through the electric field, the slower particles are deflected more than the faster.

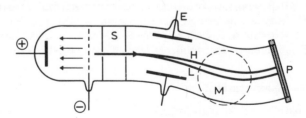

Fig. 79. Aston's mass spectrograph

This produces a scatter within each of the two streams. To overcome this, Aston used a magnetic field to bend the beams in the reverse direction. This is shown in the diagram by an indication of the magnetic poles M (the poles lie above and below the apparatus). The extent of the second deflection was adjusted to correct for the velocity scatter and yet allow the beams of heavy and light positive rays to be resolved. Finally, the positive ray streams were allowed to impinge on a photographic plate, which when developed gave a *mass spectrum* for neon. Each line in the spectrum corresponds to a different charge-to-mass ratio for the particles constituting the positive ray beam. In this way, the presence of the various isotopes may be confirmed, and by altering the strengths of the deflecting fields, the instrument can be used over any desired isotopic mass range.

Atomic weights Modern determinations of atomic weights are made using instruments such as the mass spectrograph. In these determinations, the mass of an isotope is measured relative to that of the carbon-12 isotope (the mass of the carbon-12 isotope being taken as 12 exactly). However, the smallest quantity of an element that we can handle in a chemical reaction contains millions of atoms, some of which will be of one isotope, some of another (unless the element is one of the few that have one isotope only). Consequently, the atomic weight of an element is the weighted average of the masses of the constituent isotopes. This atomic weight is referred to as the *chemical atomic weight*.

The Stability of the Nucleus

Apart from hydrogen, all atomic nuclei contain neutrons in addition to protons. It is thought that, in some way, the strong force of mutual repulsion between the protons packed within the small confines of the nucleus is overcome by the presence of neutrons, which thus help to stabilize the nucleus. It is found that for a stable nucleus the ratio of the number of protons to the number of neutrons lies between certain limits, as indicated in Figure 80. Within the shaded region, the proton–neutron ratio corresponds to that of a stable nucleus. Below the shaded area, the nucleus contains excess neutrons and is radioactive. Many nuclei of this type are created by bombardment of an element with neutrons in an atomic pile. For example,

$$^{12}_{6}C + 2(^{1}_{0}n) \rightarrow {}^{14}_{6}C \quad \text{(neutron rich)}$$
$$\downarrow$$
$$^{14}_{6}N + \beta$$

This is known as artificially induc redadioactivity.

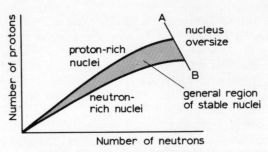

Fig. 80. Nuclear stability: neutron–proton ratio

Beyond the line AB in Figure 80, the nucleus appears to be too large to be stable, and a reduction in the overall size of the nucleus is effected by the loss of an α-particle:

$$^{238}_{92}U \rightarrow \alpha + ^{234}_{90}Th$$

Referring back to Figure 78, the uranium disintegration series can be regarded as a reduction in nuclear size by the loss of α-particles, followed by a readjustment of the proton–neutron ratio by the emission of β-particles.

Above the shaded portion of the graph in Figure 80, the nucleus contains an excess of protons. Such nuclei are produced by bombardment; for example,

$$^{27}_{13}Al + ^{4}_{2}He(\alpha) \rightarrow ^{30}_{15}P + ^{1}_{0}n$$

The phosphorus produced by this reaction is proton-rich and it is radioactive. The nucleus achieves stability by the emission of a positron (β^+) (a particle similar to a β particle but carrying a positive charge):

$$p \rightarrow n + \beta^+$$

Thus,

$$^{30}_{15}P \rightarrow ^{30}_{14}Si + \beta^+$$

Nuclear Energy

On bombardment with slow neutrons, atoms of the uranium-235 isotope capture a neutron, producing uranium-236. This isotope is unstable, but instead of regaining stability by emitting α- or β-particles, the nucleus undergoes fission, splitting into two heavy fragments:

$$^{235}_{92}U + ^{1}_{0}n \rightarrow (^{236}_{92}U) \xrightarrow{fission} ^{144}_{56}Ba + ^{90}_{36}Kr + 2(^{1}_{0}n)$$

In fact, during this process, a wide spread of fission products is formed; the barium and krypton in the above equation is just one example of many that could have been selected. In this equation, it is clear that all the original 92 protons and 144 neutrons have been accounted for, but the total masses of the products falls short of that of the initial uranium-236. This mass difference is converted into energy, which is released at the time of fission. According to Einstein's equation,
$$\text{energy} = mc^2$$
(where m is the mass and c the velocity of light), and a small mass loss can give rise to a tremendous release of energy. When fission takes place, on average two neutrons are emitted. Provided these neutrons are slowed down to the correct velocity by a moderator (graphite or heavy water) they may be captured by neighbouring uranium-235 atoms, so that the process is self-maintained and a chain reaction develops. Control of the chain reaction may be effected using neutron absorbers such as cadmium or boron alloys.

Nuclear fusion, a process in which light nuclei are made to combine, under the influence of high temperatures, to form heavier nuclei, is another type of nuclear reaction which leads to a loss of mass and the consequent release of a great deal of energy.

The Uses of Radioactive Isotopes

1. *Medical uses*. Radiations can be used to kill malignant cell tissue, while instruments pre-packed in sealed plastic containers can be sterilized by exposing the package to radiation.
2. *Tracer techniques*. The progress of an object which is rendered radioactive can be traced by picking up the emitted radiations. The use of tracer techniques is widespread, ranging from the measurement of engine wear and following the flow of fluids in pipes, to tracing the circulation of blood or measuring the uptake of nutrients by plants. Some chemical applications of this technique include the determination of the solubility of sparingly soluble salts and the investigation of the mechanism of a chemical reaction.
3. *Thickness gauging*. Both β- and γ-rays are capable of penetrating solid materials, and the amount of radiation passing through the sample becomes smaller as the thickness of the sample increases. Thus, by passing the sample under test between a source of radiation and a detector, the thickness of the material may be measured.

4. *Irradiation techniques.* These are used to sterilize food, instruments and apparatus. The radiation kills many bacteria. Irradiation of polymers can lead to the formation of more cross-links between the polymer chains, so producing a more rigid structure.
5. *Radioactivation analysis.* Many elements absorb neutrons into their nuclei when exposed to the neutron flux within an atomic pile. When this takes place, the resulting nucleus is neutron-rich and is usually radioactive. The radioactive species so formed will have a characteristic half life and the presence of the element can be demonstrated. In addition, the concentration of the element in a sample can be determined from the intensity of the radioactivity so produced, by comparison with the activity of a series of reference samples.
6. *Radioactive dating.* The half life for the overall disintegration of thorium-232 to lead-208 is 18×10^9 years. The lead-208 content of a thorium ore can be regarded as the end product of the above series, and the time needed for the proportion of lead-208 found by experiment to be present in the ore can be estimated. A correction is applied to allow for the fact that some of the lead-208 may not have been produced in this way. The results of many experiments show that the age of these minerals is about 3000 million years.

The isotope carbon-14 has a half life of 5600 years, and this isotope occurs in atmospheric carbon dioxide. While photosynthesis takes place, the radioactivity due to carbon-14 in plant tissue remains steady, but when the plant is killed, the radioactivity shows the usual decay. Thus, by measuring the activity of carbonaceous material, for example paper or wood, the age of that material may be estimated. In this way, the famous Dead Sea Scrolls have been estimated to be 1930 years old.

Appendix

ATOMIC WEIGHTS OF THE ELEMENTS

Element	Symbol	Atomic number	Atomic weight	Approximate atomic weight
Aluminium	Al	13	26·9815	27
Antimony	Sb	51	121·75	121·5
Argon	Ar	18	39·948	40
Arsenic	As	33	74·9216	75
Barium	Ba	56	137·34	137·5
Beryllium	Be	4	9·0122	9
Bismuth	Bi	83	208·98	209
Boron	B	5	10·811	11
Bromine	Br	35	79·909	80
Cadmium	Cd	48	112·4	112·5
Caesium	Cs	55	132·905	133
Calcium	Ca	20	40·08	40
Carbon	C	6	12·0111	12
Chlorine	Cl	17	35·453	35·5
Chromium	Cr	24	51·996	52
Cobalt	Co	27	58·9332	59
Copper	Cu	29	63·54	63·5
Fluorine	F	9	18·9984	19
Germanium	Ge	32	72·59	72·5
Gold	Au	79	196·967	197
Helium	He	2	4·0026	4
Hydrogen	H	1	1·00797	1
Iodine	I	53	126·9044	127
Iron	Fe	26	55·847	56
Krypton	Kr	36	83·8	84
Lead	Pb	82	207·19	207
Lithium	Li	3	6·939	7
Magnesium	Mg	12	24·312	24·5
Manganese	Mn	25	54·938	55
Mercury	Hg	80	200·59	200·5
Neon	Ne	10	20·183	20
Nickel	Ni	28	58·71	58·5
Nitrogen	N	7	14·0067	14
Oxygen	O	8	15·9994	16
Phosphorus	P	15	30·9738	31
Potassium	K	19	39·102	39
Rubidium	Rb	37	85·47	85·5
Silicon	Si	14	28·086	28
Silver	Ag	47	107·87	108
Sodium	Na	11	22·9898	23
Strontium	Sr	38	87·62	87·5
Sulphur	S	16	32·064	32
Tin	Sn	50	118·69	118·5
Xenon	Xe	54	131·30	131·5
Zinc	Zn	30	65·37	65·5

LATTICE ENERGIES (kJ mole^{-1})

Cation	Anion				Cation	Anion	
	F$^-$	Cl$^-$	Br$^-$	I$^-$		F$^-$	O^{-2}
Ag$^+$	955	901	888	884	Al^{3+}		15100
Li$^+$	1031	846	800	742	Ba^{2+}	2363	3105
Na$^+$	918	779	742	691	Ca^{2+}	2635	3510
K$^+$	813	708	679	641	Fe^{2+}	2870	3939
Rb$^+$	779	679	658	624	Mg^{2+}	2925	3825
Cs$^+$	750	654	633	603	Zn^{2+}	2816	4043
NH$_4^+$	800	675	646	607			

BOND ENERGIES (kJ mole^{-1})

Single bonds between:

	H	C	N	S	I	Br	Cl	F
H	436	415	390	339	297	364	431	566
C	415	344	293	260	239	272	327	440
Cl	431	327	201	256	214	218	243	256
F	566	440	272	297		256	256	159
O	461	352					205	184

Also: $C = C$ 612, $C = O$ 746, $C \equiv C$ 813

HYDRATION ENERGIES (kJ mole^{-1})

Li$^+$	515	Mg^{2+}	1927	Al^{3+}	4651
Na$^+$	406	Ca^{2+}	1655	Ag$^+$	461
K$^+$	323	Sr^{2+}	1487	F$^-$	536
Rb$^+$	293	Ba^{2+}	1278	Cl$^-$	406
Cs$^+$	264	Zn^{2+}	2032	Br$^-$	385
NH$_4^+$	293	H$^+$	1120	I$^-$	360

ELECTRON AFFINITIES (kJ mole^{-1})

	Energy released
F → F$^-$	−350
Cl → Cl$^-$	−366
Br → Br$^-$	−344
I → I$^-$	−317

IONIC RADII (Ångstrom units)

Li$^+$	0·68	Be^{2+}	0·30	Al^{3+}	0·48	F$^-$	1·33
Na$^+$	0·98	Mg^{2+}	0·65	Fe^{3+}	0·53	Cl$^-$	1·81
K$^+$	1·33	Ca^{2+}	0·94	Cr^{3+}	0·55	Br$^-$	1·96
Rb$^+$	1·48	Sr^{2+}	1·10	Zn^{2+}	0·70	I$^-$	2·19
Cs$^+$	1·67	Ba^{2+}	1·29	Cu^{2+}	0·69	O^{2-}	1·45
Ag$^+$	1·13	Fe^{2+}	0·75	NH$_4^+$	1·48	S^{2-}	1·90

Short reading list and source of data

For further reading

ASHMORE, P. G., *Principles of Reaction Kinetics* (Monographs for Teachers, No. 9), Royal Institute of Chemistry, 1961.

DAVIES, C. W., *Principles of Electrochemistry* (Monographs for Teachers, No. 1), Royal Institute of Chemistry, 1959.

GLASSTONE, S., *Textbook of Physical Chemistry*, 2nd ed., Macmillan, 1948.

IVES, D. J. G., *Principles of the Extraction of Metals* (Monographs for Teachers, No. 3), Royal Institute of Chemistry, 1960.

O.E.C.D., *Chemistry Today: A Guide for Teachers*, Organization for Economic Cooperation and Development, 1963.

SHARPE, A. G., *Principles of Oxidation and Reduction* (Monograph for Teachers, No. 2), Royal Institute of Chemistry, 1959.

Sources of Data

AYLWARD, G. H. and FINDLAY, T. J. V., *Chemical Data Book*, 2nd ed., Wiley, 1966.

PARSONS, R., *Electrochemical Constants*, Butterworth, 1959.

LANGE, N. A. (Ed.), *Handbook of Chemistry*, 10th ed., McGraw-Hill, 1967.

PERRY, R. H. (Ed.), *Chemical Engineers' Handbook*, 4th ed., McGraw-Hill, 1963.

Answers to exercises

Chapter 1 (p. 17)

1. (a) 90 (b) 126 (c) 124
2. (a) 0·3 g (b) 0·3 g (c) 1·2 g (d) 0·249 g (e) 0·08 g
3. (a) 0·11 (b) 5×10^{-12} (c) 0·5 (d) 1·0 (e) 1·0 (f) 0·223
4. CH, C_2H_2 5. C_6H_6 6. PCl_3
7. (a) 1·92 g (b) 9·6 g (c) 0·384 g (d) 0·0096 g
8. (a) 0·5M (b) 0·1M (c) 0·069M
9. 1M   10. $M + 2HCl = MCl_2 + H_2$ 11. 0·05M
12. $Na_2CO_3 + HCl = NaCl + NaHCO_3$

Chapter 2 (p. 51)

1. (a) 1·12 litres (b) 3·06 litres (c) 2·24 litres
2. 13 lb in^{-2} 2·6 lb in^{-2} 3. 46·0 g
4. (a) $5·55 \times 10^{-2}$ (b) 10^{-2} (c) $4·24 \times 10^{-2}$ 6. 17·2 g
7. 738 litres   8. 4 litres   9. $C_2H_4Br_2$ 10. 14·6 cm
11. C_2H_4, 58·3% CH_4; 41·7% C_2H_4 12. 79000 cm s^{-1}
13. 43·7 g   14. 91·4 g   15. (a) 0·0023 (b) 991 mm
16. 53·4 g   17. 120 g (associated into double molecules)
18. 0·55 mole
19. $1·35 \times 10^{-5}$. Mole fractions: H_2O 0·162, CO_2 0·344, N_2 0·494
20. 1·16 atm, 0·232 atm   21. 28·04

Chapter 3 (p. 98)

1. (a) 80°C (b) 71°C, 1·5 atm   2. 55·3 mm   4. 600 cm^3
5. 0·0213   6. 11·5°C, 65% by weight p-bromotoluene
7. 18·4 cm   8. 13·5° to 101°C   9. 20·7   10. 176·4
11. 18·5% and 66% by weight isobutyric acid   12. 472·9 mm
13. 100·15°C   14. 78·68°C   15. 239 (molecules associated)
16. 2·62 atm   17. $-1·86$°C   18. 20 g   19. 8·99 g l^{-1}
20. 49, 117·6
21. 59·75 cm^3; 6·3% nitrogen, 10·0% oxygen, 83·7% carbon dioxide
22. 91·7   23. $-0·32$°C   25. C_7H_6O 26. 255
27. 3·45 atm

Chapter 4 (p. 142)

1. C_v 21 J mole^{-1}K^{-1}, C_p 29·3 J mole^{-1}K^{-1}, $\gamma = 1·39$
2. 7·0° and 5·4°C   3. 1280 kJ   5. $\Delta H = 114$ kJ
6. C_2H_4 $\Delta H = 19$ kJ; C_2H_6 $\Delta H = -107$ kJ   7. 2·27 kJ
9. $\Delta H = -496$ kJ
10. $\Delta G = -24·9$ kJ at 300 K, $\Delta G = 18·7$ kJ at 500 K
11. $\Delta H = -139$ kJ
12. -964 kJ mole^{-1} (a) 163 kJ mole^{-1} (b) -414 kJ mole^{-1}
13. -660 kJ mole^{-1} 14. $\Delta H_f = +573$ kJ mole^{-1}
15. -305 kJ mole^{-1} 16. $E_{C-Cl} = 329$ kJ mole^{-1}
17. $E_{C-C} = 332$ kJ mole^{-1}
18. $\Delta H_f = 217$ kJ mole^{-1} in accordance with H—C≡C—H
19. $E_{C-Cl} = 293$ kJ mole^{-1} $E_{C-Br} = 268$ kJ mole^{-1} 20. 87·2 kJ
21. (a) 421 kJ (b) -127 kJ reduction feasible

Chapter 5 (p. 174)

1. 45·9 g   2. 0·014   3. 0·074 atm   4. 27·1 g
5. (a) 16 mole H_2, 4 mole N_2 (b) 20 moles (c) 18 moles
   (d) 13 mole H_2, 3 mole N_2, 2 mole NH_3 (e) $0·0006 V^2$
6. 833 1 mole^{-1} 7. 1·19 atm   8. 73·7%   9. 0·8 g
10. 19·5%, 84%, endothermic   11. 0·39 or 39%   12. 94·1
13. 100 cm^3, 4·44 g   15. 103 kJ
16. $2·4 \times 10^{-8}$ 1 mole^{-1} s^{-1}

Chapter 6 (p. 222)

1. 0·287 g   2. A 0·695, B 0·820, 139 ohm^{-1} cm^2
3. 0·426 g, 0·48 g, 134·4 cm^3, 1·775 g (i.e. all the chlorine in the solution)   4. 200 ohm^{-1} cm^2 5. 0·976
6. $2·33 \times 10^{-3}$ g l^{-1} 7. 347·3, 0·032 or 3·2%   8. 0·058%
9. $1·36 \times 10^{-3}$ g l^{-1} 10. 12·7, 0·7   11. 0·14, 0·44%
12. 11·1   13. $3·98 \times 10^{-5}$ mole l^{-1} 14. $3·16 \times 10^{-6}$ mole l^{-1}
15. 5·72   16. $2·5 \times 10^{-11}$ mole^2 l^{-2} 17. 1×10^{-10} mole l^{-1}
18. 10^{-15} mole l^{-1}, 10^{-25} mole l^{-1} 19. 0·009 mole
20. 0·00958 g, $1·21 \times 10^{-5}$ g   21. 0·44 V   22. 0·35 V
23. $-0·88$ V

Table of Common Logarithms

N	0	1	2	3	4	5	6	7	8	9	Proportional Parts								
											1	2	3	4	5	6	7	8	9
10	0000	0043	0086	0128	0170	0212	0253	0294	0334	0374	4	8	12	17	21	25	29	33	37
11	0414	0453	0492	0531	0569	0607	0645	0682	0719	0755	4	8	11	15	19	23	26	30	34
12	0792	0828	0864	0899	0934	0969	1004	1038	1072	1106	3	7	10	14	17	21	24	28	31
13	1139	1173	1206	1239	1271	1303	1335	1367	1399	1430	3	6	10	13	16	19	23	26	29
14	1461	1492	1523	1553	1584	1614	1644	1673	1703	1732	3	6	9	12	15	18	21	24	27
15	1761	1790	1818	1847	1875	1903	1931	1959	1987	2014	3	6	8	11	14	17	20	22	25
16	2041	2068	2095	2122	2148	2175	2201	2227	2253	2279	3	5	8	11	13	16	18	21	24
17	2304	2330	2355	2380	2405	2430	2455	2480	2504	2529	2	5	7	10	12	15	17	20	22
18	2553	2577	2601	2625	2648	2672	2695	2718	2742	2765	2	5	7	9	12	14	16	19	21
19	2788	2810	2833	2856	2878	2900	2923	2945	2967	2989	2	4	7	9	11	13	16	18	20
20	3010	3032	3054	3075	3096	3118	3139	3160	3181	3201	2	4	6	8	11	13	15	17	19
21	3222	3243	3263	3284	3304	3324	3345	3365	3385	3404	2	4	6	8	10	12	14	16	18
22	3424	3444	3464	3483	3502	3522	3541	3560	3579	3598	2	4	6	8	10	12	14	15	17
23	3617	3636	3655	3674	3692	3711	3729	3747	3766	3784	2	4	6	7	9	11	13	15	17
24	3802	3820	3838	3856	3874	3892	3909	3927	3945	3962	2	4	5	7	9	11	12	14	16
25	3979	3997	4014	4031	4048	4065	4082	4099	4116	4133	2	3	5	7	9	10	12	14	15
26	4150	4166	4183	4200	4216	4232	4249	4265	4281	4298	2	3	5	7	8	10	11	13	15
27	4314	4330	4346	4362	4378	4393	4409	4425	4440	4456	2	3	5	6	8	9	11	13	14
28	4472	4487	4502	4518	4533	4548	4564	4579	4594	4609	2	3	5	6	8	9	11	12	14
29	4624	4639	4654	4669	4683	4698	4713	4728	4742	4757	1	3	4	6	7	9	10	12	13
30	4771	4786	4800	4814	4829	4843	4857	4871	4886	4900	1	3	4	6	7	9	10	11	13
31	4914	4928	4942	4955	4969	4983	4997	5011	5024	5038	1	3	4	6	7	8	10	11	12
32	5051	5065	5079	5092	5105	5119	5132	5145	5159	5172	1	3	4	5	7	8	9	11	12
33	5185	5198	5211	5224	5237	5250	5263	5276	5289	5302	1	3	4	5	6	8	9	10	12
34	5315	5328	5340	5353	5366	5378	5391	5403	5416	5428	1	3	4	5	6	8	9	10	11
N	0	1	2	3	4	5	6	7	8	9	1	2	3	4	5	6	7	8	9
											Proportional Parts								

Table of Common Logarithms—(contd)

N	0	1	2	3	4	5	6	7	8	9	Proportional Parts								
											1	2	3	4	5	6	7	8	9
35	5441	5453	5465	5478	5490	5502	5514	5527	5539	5551	1	2	4	5	6	7	9	10	11
36	5563	5575	5587	5599	5611	5623	5635	5647	5658	5670	1	2	4	5	6	7	8	10	11
37	5682	5694	5705	5717	5729	5740	5752	5763	5775	5786	1	2	3	5	6	7	8	9	10
38	5798	5809	5821	5832	5843	5855	5866	5877	5888	5899	1	2	3	5	6	7	8	9	10
39	5911	5922	5933	5944	5955	5966	5977	5988	5999	6010	1	2	3	4	5	7	8	9	10
40	6021	6031	6042	6053	6064	6075	6085	6096	6107	6117	1	2	3	4	5	6	8	9	10
41	6128	6138	6149	6160	6170	6180	6191	6201	6212	6222	1	2	3	4	5	6	7	8	9
42	6232	6243	6253	6263	6274	6284	6294	6304	6314	6325	1	2	3	4	5	6	7	8	9
43	6335	6345	6355	6365	6375	6385	6395	6405	6415	6425	1	2	3	4	5	6	7	8	9
44	6435	6444	6454	6464	6474	6484	6493	6503	6513	6522	1	2	3	4	5	6	7	8	9
45	6532	6542	6551	6561	6571	6580	6590	6599	6609	6618	1	2	3	4	5	6	7	8	9
46	6628	6637	6646	6656	6665	6675	6684	6693	6702	6712	1	2	3	4	5	6	7	7	8
47	6721	6730	6739	6749	6758	6767	6776	6785	6794	6803	1	2	3	4	5	5	6	7	8
48	6812	6821	6830	6839	6848	6857	6866	6875	6884	6893	1	2	3	4	4	5	6	7	8
49	6902	6911	6920	6928	6937	6946	6955	6964	6972	6981	1	2	3	4	4	5	6	7	8
50	6990	6998	7007	7016	7024	7033	7042	7050	7059	7067	1	2	3	3	4	5	6	7	8
51	7076	7084	7093	7101	7110	7118	7126	7135	7143	7152	1	2	3	3	4	5	6	7	8
52	7160	7168	7177	7185	7193	7202	7210	7218	7226	7235	1	2	2	3	4	5	6	7	7
53	7243	7251	7259	7267	7275	7284	7292	7300	7308	7316	1	2	2	3	4	5	6	6	7
54	7324	7332	7340	7348	7356	7364	7372	7380	7388	7396	1	2	2	3	4	5	6	6	7
55	7404	7412	7419	7427	7435	7443	7451	7459	7466	7474	1	2	2	3	4	5	5	6	7
56	7482	7490	7497	7505	7513	7520	7528	7536	7543	7551	1	2	2	3	4	5	5	6	7
57	7559	7566	7574	7582	7589	7597	7604	7612	7619	7627	1	2	2	3	4	5	5	6	7
58	7634	7642	7649	7657	7664	7672	7679	7686	7694	7701	1	1	2	3	4	4	5	6	7
59	7709	7716	7723	7731	7738	7745	7752	7760	7767	7774	1	1	2	3	4	4	5	6	7
N	0	1	2	3	4	5	6	7	8	9	1	2	3	4	5	6	7	8	9
											Proportional Parts								

Table of Common Logarithms—(contd)

N	0	1	2	3	4	5	6	7	8	9	\multicolumn{9}{c}{Proportional Parts}
											1	2	3	4	5	6	7	8	9
60	7782	7789	7796	7803	7810	7818	7825	7832	7839	7846	1	1	2	3	4	4	5	6	6
61	7853	7860	7868	7875	7882	7889	7896	7903	7910	7917	1	1	2	3	4	4	5	6	6
62	7924	7931	7938	7945	7952	7959	7966	7973	7980	7987	1	1	2	3	4	4	5	6	6
63	7993	8000	8007	8014	8021	8028	8035	8041	8048	8055	1	1	2	3	3	4	5	5	6
64	8062	8069	8075	8082	8089	8096	8102	8109	8116	8122	1	1	2	3	3	4	5	5	6
65	8129	8136	8142	8149	8156	8162	8169	8176	8182	8189	1	1	2	3	3	4	5	5	6
66	8195	8202	8209	8215	8222	8228	8235	8241	8248	8254	1	1	2	3	3	4	5	5	6
67	8261	8267	8274	8280	8287	8293	8299	8306	8312	8319	1	1	2	3	3	4	5	5	6
68	8325	8331	8338	8344	8351	8357	8363	8370	8376	8382	1	1	2	3	3	4	4	5	6
69	8388	8395	8401	8407	8414	8420	8426	8432	8439	8445	1	1	2	2	3	4	4	5	6
70	8451	8457	8463	8470	8476	8482	8488	8494	8500	8506	1	1	2	2	3	4	4	5	6
71	8513	8519	8525	8531	8537	8543	8549	8555	8561	8567	1	1	2	2	3	4	4	5	5
72	8573	8579	8585	8591	8597	8603	8609	8615	8621	8627	1	1	2	2	3	4	4	5	5
73	8633	8639	8645	8651	8657	8663	8669	8675	8681	8686	1	1	2	2	3	4	4	5	5
74	8692	8698	8704	8710	8716	8722	8727	8733	8739	8745	1	1	2	2	3	4	4	5	5
75	8751	8756	8762	8768	8774	8779	8785	8791	8797	8802	1	1	2	2	3	3	4	5	5
76	8808	8814	8820	8825	8831	8837	8842	8848	8854	8859	1	1	2	2	3	3	4	5	5
77	8865	8871	8876	8882	8887	8893	8899	8904	8910	8915	1	1	2	2	3	3	4	4	5
78	8921	8927	8932	8938	8943	8949	8954	8960	8965	8971	1	1	2	2	3	3	4	4	5
79	8976	8982	8987	8993	8998	9004	9009	9015	9020	9025	1	1	2	2	3	3	4	4	5
80	9031	9036	9042	9047	9053	9058	9063	9069	9074	9079	1	1	2	2	3	3	4	4	5
81	9085	9090	9096	9101	9106	9112	9117	9122	9128	9133	1	1	2	2	3	3	4	4	5
82	9138	9143	9149	9154	9159	9165	9170	9175	9180	9186	1	1	2	2	3	3	4	4	5
83	9191	9196	9201	9206	9212	9217	9222	9227	9232	9238	1	1	2	2	3	3	4	4	5
84	9243	9248	9253	9258	9263	9269	9274	9279	9284	9289	1	1	2	2	3	3	4	4	5
N	0	1	2	3	4	5	6	7	8	9	\multicolumn{9}{c}{Proportional Parts}
											1	2	3	4	5	6	7	8	9

Table of Common Logarithms—(contd)

N	0	1	2	3	4	5	6	7	8	9	Proportional Parts								
											1	2	3	4	5	6	7	8	9
85	9294	9299	9304	9309	9315	9320	9325	9330	9335	9340	1	1	2	2	3	3	4	4	5
86	9345	9350	9355	9360	9365	9370	9375	9380	9385	9390	1	1	2	2	3	3	4	4	5
87	9395	9400	9405	9410	9415	9420	9425	9430	9435	9440	0	1	1	2	2	3	3	4	4
88	9445	9450	9455	9460	9465	9469	9474	9479	9484	9489	0	1	1	2	2	3	3	4	4
89	9494	9499	9504	9509	9513	9518	9523	9528	9533	9538	0	1	1	2	2	3	3	4	4
90	9542	9547	9552	9557	9562	9566	9571	9576	9581	9586	0	1	1	2	2	3	3	4	4
91	9590	9595	9600	9605	9609	9614	9619	9624	9628	9633	0	1	1	2	2	3	3	4	4
92	9638	9643	9647	9652	9657	9661	9666	9671	9675	9680	0	1	1	2	2	3	3	4	4
93	9685	9689	9694	9699	9703	9708	9713	9717	9722	9727	0	1	1	2	2	3	3	4	4
94	9731	9736	9741	9745	9750	9754	9759	9763	9768	9773	0	1	1	2	2	3	3	4	4
95	9777	9782	9786	9791	9795	9800	9805	9809	9814	9818	0	1	1	2	2	3	3	4	4
96	9823	9827	9832	9836	9841	9845	9850	9854	9859	9863	0	1	1	2	2	3	3	4	4
97	9868	9872	9877	9881	9886	9890	9894	9899	9903	9908	0	1	1	2	2	3	3	4	4
98	9912	9917	9921	9926	9930	9934	9939	9943	9948	9952	0	1	1	2	2	3	3	4	4
99	9956	9961	9965	9969	9974	9978	9983	9987	9991	9996	0	1	1	2	2	3	3	3	4
N	0	1	2	3	4	5	6	7	8	9	1	2	3	4	5	6	7	8	9
											Proportional Parts								

INDEX

Absolute temperature, 22
Absolute zero, 22
Absorption coefficient, 60, 63
Acetamide–benzoic acid system, 82
Acetic acid (conductivity), 198
Acids, 204, 205
   pH of, 207–210
   strengths of, 205, 206
   weak, 199, 204–209
Activated complex, 164
Activation energy, 165–167, 193
Active mass, 147, 157, 203
Activity coefficient, 203, 204
Additivity of bond energies, 127
Adsorption, 174
Alloy, 82
α-ray, 232
Ammonia as a solvent, 205
Andrew's experiments, 40–41
Anion, 190
Anode, 189
Apparent degree of ionization, 200
Aprotic solvent, 204
Arrhenius, S., 165–167
Arsenic (III) sulphide colloid, 228
Artificially induced radioactivity, 239
Association, 97, 160
Atomic energy, 241
Atomicity, 12, 108
Atomic theory, 12
Atomic weights, 13, 239
   table of, 243
Atomization, heats of, 114, 127
Atoms, 12
Attraction of molecules, 37
Auto-catalysis, 173
Avogadro's hypothesis, 12, 13, 31, 44
Avogadro's number, 14
Azeotropes, 73, 74

Balanced reaction, 147
Bases, 204, 205
   pH of, 207–210
   strengths of, 205, 206
   weak, 199, 204, 205
Beckmann's method, 88–93
Beckmann's thermometer, 89, 90
Benzoic acid–acetamide system, 82
Berkeley and Hartley's method, 95
β-rays, 232
Boiling point, 55
   composition curves, 65, 66
   elevation, 86–90
   maximum, 72
   minimum, 73, 74
Bombardment of nucleus, 239
Bomb calorimeter, 113, 114
Bond, covalent, 124–134
   dissociation energy, 126
Bond energy, 115, 124–132
   and electronegativity, 132, 133
   and reactivity, 129–133
   determination of, 125
   table of values, 244
Bond
   fission, 163, 164
   formation, 164
   ionic, 115–118
   types, 115
Born equation, 123
Born–Haber cycle, 116, 118
Boyle's law, 20
Bredig's arc method, 230
Brønsted–Lowry theory, 204
Brownian motion, 227
Bubble cap column, 67, 69
Buffer solution, 215–218

Calomel electrode, 183
Calorimeter, bomb, 113, 114
Carbon dioxide isothermals, 40

Catalysis, 172–174
  and equilibrium, 156
Cataphoresis, 228
Cathode, 189
Cation, 190
Cell
  concentration, 187
  conductivity, 194, 195
  constant, 195, 196
  diagrams, 180, 181
  electrolytic, 190
  fuel, 188
  galvanic, 177, 178
  half, 180, 181
  voltaic, 177, 178
Charles' law, 22
Chemical atomic weights, 239
Chloroform, vapour pressure, 55, 56
Closed system, 154
Coagulation of sols, 229
Colligative properties, 84–98
Collision geometry, 164
Collision theory of reaction rates, 171
Colloid
  arsenic (III) sulphide, 228
  iron (III) hydroxide, 227
  irreversible, 227
  lyophobic, 227
  lyophilic, 226
  mill, 230
  reversible, 227
Colloidal particles (size of), 226
Colloidal solutions, 225–231
  coagulation of, 229
  electrical properties, 227, 228
  optical properties, 227
  preparation of, 229–231
  stability of, 228, 229
Colloidal state, 225–231
Colour of indicators, 214
Combustion, heat of, 110
Common ion effect, 220, 221
Concentration cell, 187
Concentration of solutions, 56, 57
Conduction
  electrolytic, 189
  metallic, 189

Conductivity
  and dilution, 198
  cell, 195, 196
  electrolytic, 195
  equivalent, 197–201
  ionic, 200
  molar, 197
  of electrolytes, 194–201
  specific, 194–197
  water, 206, 207
Conjugate solutions, 75
Conservation of energy, 12, 106
Consolute temperature, 76
Constant
  cell, 195, 196
  dissociation, 202
  equilibrium, 148
  velocity, 148, 165, 166
Contact process, 156, 157
Continuity of state, 41
Copper sulphate, electrolysis, 194
Coulomb, 190
Coulometer, 178
Covalent bonding, 124–134
o-cresol water system, 75
Critical
  pressure, 41
  solution temperature, 75, 76
  temperature, 41
  volume, 41
Crystallization, 80, 81
Curie (of radiation), 235

Dalton's
  atomic theory, 12
  extension to Henry's law, 63
  law of partial pressures, 27
Daniell cell, 178, 181
Dating, radioactive, 242
Debye–Hückel theory, 203
Decay curve, 233, 234
Decay, radioactive, 232–234
Decomposition voltage, 192, 193
Degree of dissociation, 158
Degree of ionization, 199–202
Deliquescence, 84
Delocalization of electrons, 129

Deposition potential, 192, 193
Depression of freezing point, 90–93
   Beckmann's method, 92
   constant, 91
   Rast's micro method, 92–93
Depression, molar, 91
Diffusion, 33–36
Dilute solutions, 84–98
Dilution
   and conductivity, 198
   infinite, 199
   law, 202
Dipole moment, 133
Discharge of ions, 192, 193
Disintegration, radioactive, 233
Disintegration series, 237
Disorder, 135
Disperse phase, 226
Dispersion medium, 226
Displacement of equilibrium, 153–156
Dissociation, 97, 98
   constant, 202
   degree of, 158, 199–202, 210
   energy of a bond, 126
   thermal, 158
Distillation
   columns, 67
   extractive, 74
   steam, 27
   theory of, 67–69
Distribution
   coefficient, 159
   law, 159
   of molecular energies, 162
   of molecular velocities, 33
Dumas' method, 46, 47
Dynamic equilibrium, 147

Efflorescence, 84
Effusiometer, 35
Effusion, 33–36
Electrochemical equivalent, 191
Electrochemistry, 177–199
Electrode
   calomel, 183
   carbon, 194
   glass, 188
   mercury, 193
   platinum, 193
   potentials, 179–183
   standard hydrogen, 180
   used to measure pH, 187, 188
Electrolysis, 189–194
   Faraday's laws, 190
Electrolyte
   conductivity of, 194–201
   strong, 199, 203, 204
   weak, 199, 203, 204
Electrolytic
   cell, 190
   conduction, 189
   conductivity, 195
Electron affinity, 116, 117, 121
   table of, 244
Electronegativity and bond energy, 132, 133
Electron flow, 180, 189, 190
Electro-osmosis, 228
Electrophoresis, 228
Elevation of boiling point, 86–90
   constant, 87
   molar, 87
Ellingham diagram, 140, 141
Empirical formula, 15
Endothermic compound, 113
Endothermic reaction, 109, 113, 139, 163
Energy
   activation, 165–167, 193
   and bonding, 115
   bond, 115, 124–132
      determination of, 125
      dissociation, 126
      table of, 244
   changes, 102–144
   forms of, 102
   free, 135–139, 184
   heat, 102, 103
   hydration, 119
      table of, 244
   internal, 102, 104–106
   kinetic, 30, 31, 102, 104, 105
   lattice, 116–122
   of gas particles, 30, 31

Energy (*cont.*)
  potential, 102
  rotational, 102, 106
  solvation, 119
  translational, 102
  units of, 102
  vibrational, 102, 106
Enthalpy, 109
Entropy, 135–141
Equation, ideal gas, 23
Equation of state, 38, 39
Equilibrium, 146–161
  and concentration, 148, 154
  constant, 148
  displacement, 153–156
  dynamic, 147
  effect of catalysts on, 156
  effect of pressure on, 155
  effect of temperature on, 155
  freezing, 150
  heterogeneous, 148–157
  homogeneous, 148–157
  in immiscible solvents, 159
  ionic, 177
  reactions of industrial importance, 156, 157
Equivalent conductivity, 197, 198
Equivalent weight of ion, 191, 197
Eutectic, 82, 83
Exothermic reaction, 109, 138, 139, 163
External work, 106–108
Extractive distillation, 74

Factors governing chemical reaction, 134–136
Factors governing size of lattice energy, 121–122
Faraday constant, 191
Faraday's laws of electrolysis 190, 191
First law of thermodynamics, 105, 106
First order reaction, 170
Fission of nucleus, 241
Fluorination by alkali fluorides, 119, 120
Forces of attraction, 37, 54

Formation, heat of, 110, 120, 121
Fractional crystallization, 80
Fractional distillation, 68, 69
Free energy, 135–139, 184
Freezing of an equilibrium, 150
Freezing point depression, 90–93
Fuel cell, 188
Fusion mixture, 82
Fusion, nuclear, 241

Galvanic cell, 177, 178
γ-ray, 232
Gas constant (R), 24
Gases, 20–51
  density by microbalance, 49, 50
  liquefaction, 40, 43
  molar heat capacity, 105, 106
  relative density, 43, 44
  vapour density, 43, 44
Gas laws, 20–23
  calculations on, 25–27
Gaseous state, 20
Gay Lussac's law, 22
Gay Lussac's law of combining volumes, 13
Geiger-Müller counter, 233
Gel, 226, 227
Gelatin colloid, 227, 231
Geometry, collision, 163, 164
Glass electrode, 188
Glycerol–guaiacol system, 77
Gold number, 229
Graham's law, 34
Group displacement law, 236, 237

Haber process, 156
Half-life times, 170, 171, 233
Half reactions, 179, 180
Halogen replacement reactions, 119, 120
Heat capacity, molar, 103, 106–108
Heat capacity of gases (molar)
  at constant pressure, 106
  at constant volume, 105, 106
  diatomic, 108
  monatomic, 108
  polyatomic, 106
  ratio of, 108

Heat
 content, 109
 energy, 102, 103
 of atomization, 114, 127
 of combustion, 110
  determination of, 113, 114
 of formation, 110, 116
  calculation of, 120, 121
 of hydrogenation, 113, 115
 of neutralization, 110, 206
 of reaction, 109
Helmholtz double layer, 228, 229
Henry's law, 60
Hess's law, 110, 111, 116
Heterogeneous catalysis, 172
Heterogeneous equilibrium, 148, 157
H.E.T.P., 69
Homogeneous catalysis, 172
Homogeneous equilibrium, 148, 157
Hydrates
 solubility curves, 79, 80
 vapour pressure of, 83, 84
Hydration energy, 119
 table of, 244
Hydrogen
 electrode, 180
 iodine reaction, 149
 overvoltage of, 193
Hydrogenation, heat of, 113, 115
Hydrolysis of salts, 211, 212
Hypothetical compounds, heat of formation of, 120, 121

Ideal gas, 23
 and osmotic pressure, 96, 97
 isothermals, 42
Ideal solution, 64
Indicators
 acid/alkali, 212–215
 pH range of, 214
 theory of, 214
Induced radioactivity, 239
Industrially important equilibrium reactions, 156, 157
Infinite dilution, 199
Intermediate compound theory, 173

Intermolecular attraction, 37, 54
Internal energy, 102, 104–106
Internal pressure, 39
Inversion temperature, 37, 54
Ionic
 attraction, 121, 122
 bonding, 115–118
 compound, melting points, 118
 conductances, 200
 equilibria, 177
 product of water, 206, 207
 radii, 244
 sizes and lattice energy, 122
 theory, 203
Ionization
 apparent degree of, 200
 degree of, 199, 202
 self, 205, 206
Ions
 discharge of, 192–194
 electrochemical equivalent of, 191
Iron (III) hydroxide colloid, 227, 228
Irradiation, 242
Irreversible colloid, 237
Iso-electric point, 237
Isoteniscope, 58, 59
Isothermals, 20, 21
 and Andrews' experiment, 40, 41
 ideal gas, 42
Isotopes, 237
i, Van't Hoff factor, 98

Joule (of energy), 102, 103
Joule–Thomson effect, 37, 43

Kelvin (degrees), 22
Kinetic energies, 31, 102
Kinetics of reaction, 136, 162–172
Kinetic theory of gases, 28–33, 37, 38
Kohlrausch's law, 200

Lattice energy, 116–119
 calculation of, 122, 123
 table of, 244

Law
  Boyle's, 20
  Charles', 22
  Faraday's, 190, 191
  Gay-Lussac's, 13, 22
  Graham's, 34
  group displacement, 236, 237
  Kohlrausch, 200
  Maxwell's distribution, 33, 162, 163
  of combining volumes, 13
  of conservation of energy, 12, 106
  of conservation of mass, 11
  of constant composition, 11
  of mass action, 147
  of multiple proportions, 12
  of osmotic pressure, 96
  of partial pressures, 27
  of reciprocal proportions, 12
  Ohm's, 177
  Ostwald's dilution, 202
Le Chatelier's principle, 59, 153
Light scattering by colloids, 227
Liquefaction of gases, 40, 43
Liquid mixtures, 64–77
Liquid state, 20, 54
Liquids
  and solutions, 54–98
  boiling point of, 55
  partially miscible, 74–76
Liquidus curve, 81
Lowering of vapour pressure, 85, 86, 90, 91
Lyophilic colloids, 227
Lyophobic colloids, 226

Madelung constant, 123
Mass action law, 147
Mass spectrograph, 238
Mathematics of radioactive decay, 234–235
Maximum boiling point, 72
Maxwell's distribution law, 33, 162, 163
Mean free path, 36
Mechanism of catalysis, 173–174
Mechanism of reaction, 163, 168, 169

Melting point of ionic compounds, 118, 119
Metallic conduction, 189
Metal oxide reduction, 139–141
Metal sulphides, solubility product, 222
Methyl orange, 212, 213
Microbalance, 49, 50
Minimum boiling point, 73, 74
Mobile equilibrium, 153
Molality of solutions, 16, 57
Molar
  conductivity, 197
  depression, 91
  elevation, 87
  gas constant, 24
  heat capacity, 103, 106–108
Molarity of solutions, 16, 57
Mole, 14, 15
Molecular formula, 15
Molecularity of a reaction, 171
Molecular theory, 12, 13
Molecular velocities, 32
  determination of, 32
  distribution of, 33, 162, 163
Molecular weights, 13, 14
  determination by density measurements, 44–50
  determination by effusiometer, 36
Mole fraction, 28
Monatomic gas, internal energy of, 104

Natural disintegration series, 237
Negative catalysis, 172
Nernst equation, 186
Neutralization, heat of, 110, 206
Neutron–proton ratio, 240
Neutron-rich nucleus, 239
Non-standard states, 185
Non-stoichiometric compounds, 12
Normality of solutions, 57
Nuclear
  energy, 240
  fission, 241
  fusion, 241
  reactions, 236, 240
Nucleus, stability of, 239

Ohm's law, 177
Open system, 154
Optimum temperature, 156
Order and disorder, 135
Order or a reaction, 168–170
Osmosis, 93–97
Osmotic pressure, 94–97
  laws of, 96
Ostwald's dilution law, 202
Ostwald's process, 157
Overvoltage, 193
Oxidation, 179, 183, 184
Oxides, reduction of metal, 139–141
Oxidizing power, 183, 184, 186, 187
Oxygen, overvoltage, 193

Partial miscibilty, 74–76
Partial pressure, 27
Partition coefficient, 159
Partition law, 159
Pauling's electronegativity values, 133
Peptization, 226, 230
Pfeffer's osmotic experiments, 95, 96
Phenolphthalein, 212, 213
pH, 207–215
  meter, 188
  of buffer solutions, 216–218
Phosphorus pentachloride thermal dissociation, 150, 151
Physical chemistry
  laws, 11, 12
  scope, 11
Polar bond, 133
Polarizability of ions, 124
Potassium chloride solution, electrolytic conductivity, 198
Potential
  energy, 102
  redox, 181, 182
  standard electrode, 179, 180
Preferential discharge of ions, 193, 194
Pressure
  critical, 41
  effect on equilibrium, 155
  internal, 39

of water vapour, 27
partial, 27
vapour, 54–56
Promoter (of a catalyst), 173
Protective action, 229
Proton
  acceptor, 204
  donor, 204
  -rich nucleus, 240
  solvated, 204
Protophilic solvent, 204

Qualitative analysis and solubility products, 221, 222

Radiation (types), 232
Radii of ions, 244
Radioactivation analysis, 242
Radioactive dating, 242
Radioactive isotopes, 237
  uses of, 242
Radioactivity, 232–242
Radon, 233, 234
Raoult's law, 64, 65, 69–73, 85
  deviations from, 70–73
Rast's micro method, 92, 93
Rate determining step, 171
Rate of reaction, 147
Ratio of heat capacities of gases, 108
Rays (α, β, γ), 232
Reaction
  collision theory, 171
  endothermic, 109, 113, 139, 164
  equilibrium, 146–161
  exothermic, 109, 138, 139, 164
  factors governing, 134–141
  first order, 170, 171
  heats of, 109
  kinetics of, 136, 162–172
  mechanism, 163, 168, 169
  molecularity of, 171
  order of, 168, 169
  rates, 147
  reversible, 146
Reactivity and bond energy, 129–132
Real gas behaviour, 37, 38

Real gases and van der Waals equation, 38
Redox potential, 181, 182
Reducing power, 183, 184
Reduction, 179, 183, 184
 of metal oxides, 139–141
Regnault's method, 44
Relative density of a gas, 43, 44
Relative lowering of vapour pressure, 85, 86, 90, 91
Reversible reactions, 146
R (molar gas constant), 24
Root mean square velocity, 32
Rotational energy, 102, 106

Salt hydrates, vapour pressure, 83, 84
Salting out effect, 64
Salts, hydrolysis of, 211, 214
Saturated solution, 76
Saturated vapour pressure, 55
Scaler, 233
Self-ionization, 205, 206
Semi-permeable membrane, 93–96
Series, disintegration, 237
Silver halides
 lattice energies, 119, 244
 solubilities, 119
SI units, 24, 102–104
Sodium chloride, electrolysis of, 193
Sodium halides, melting points, 119
Solid solution, 82
Solid state, 20
Solidus curve, 81, 82
Sols, 225
Solubility, 78
 and solubility product, 220
 curves, 79
 in polar solvents, 119
 of gases, 60–64
 of gas mixtures, 63
 of silver halides, 119
 of sparingly soluble gases, 201
Solubility product, 219–222
 and qualitative analysis, 221
 of metal sulphides, 222
Solute, 56

Solutions, 54–98, 225
 buffer, 215, 216–218
 concentrations of, 56, 57
 conjugate, 75
 definition of, 56
 ideal, 64
 molality, 57
 molarity, 16, 57
 normality, 57
 of gases in gases, 57, 58
 of gases in liquids, 59–62
 of liquids in liquids, 64–76
 of solids in liquids, 76
 of solids in solids, 81, 82
 pH of salt, 211, 214
 properties of dilute, 84–98
 saturated, 76–78
 supersaturated, 78
 types of, 57
Solvation energy, 119
Solvent, 56
 extraction, 161, 162
Sparingly soluble salts, 201
Specific
 conductivity, 194–196
 heat capacity, 103
 resistance, 194
Stability of nucleus, 239
Standard
 electrode potential, 179, 180
 hydrogen electrode, 180
 states, 142
Starch colloid, 227, 231
State, continuity of, 41
States of matter, 20, 54
Steam distillation, 27
Stoichiometry, 12
Strength of acids, 205, 206
Strong acids, 199, 204, 205
Strong bases, 199, 204, 205
Strong electrolyte, 199, 203, 204
Structural formula, 15, 16
Structure and bond energy, 128, 129
Supersaturated solution, 78
Surface action theory (of catalysis), 175
Surface tension, 54

Suspensions, 225

Temperature
　absolute, 22
　and equilibrium, 155
　and molecular velocities, 31, 32
　consolute, 76
　critical, 41
　critical solution, 75, 76
　inversion, 37, 43
　Kelvin, 22
　optimum, 156
Theoretical plate, 69
Theory of catalysis, 172–174
Theory of indicators, 212
Thermal decomposition, 150, 151
Thermal dissociation, 158
Thermochemical equations, 109–113
Thermodynamics, first law, 105, 106
Thickness gauging, 242
Thixotropic system, 227
Titration curves, 212–214
Tracer techniques, 242
Transition state, 164
Translational energy, 102
Transpiration method, 57
Trimethylpyridine–water system, 77
Tyndall cone, 227

Ultra-centrifuge, 231
Ultra-microscope, 227
Units, SI, 102–104
Uranium
　decay, 236, 237
　fission, 240, 241
　isotopes, 237

Van der Waals equation, 38, 42
Van't Hoff
　and osmotic pressure laws, 96
　factor, i, 98

Vapour density, 43, 44
　measurement of, 44–50
Vapour pressure, 54–56
　and temperature, 56, 57
　composition curves, 66
　determination of, 57, 58
　lowering of, 85–91
　of liquid mixtures, 64–72
　of salt hydrates, 83, 84
　relative lowering of, 85, 86, 90, 91
Velocity constant, 148, 165, 166
Velocity of molecules, 31, 32
　distribution law, 33
　mean square, 30
　root mean square, 32
Vibrational energy, 102, 106
Victor Meyer's method, 47, 48
Voltage, decomposition, 192
Voltaic cell, 177, 178
Volume percentage, 57

Water
　as a solvent, 205
　conductivity, 206
　ionic product, 206, 207
　pH of, 208
Weak acids, 199, 204, 205
　pH of, 207–209
Weak bases, 199, 204, 205
　pH of, 207–210
Weak electrolyte, 199, 203, 204
Work, 103
　external, 106–108
　of expansion, 107

Zero of temperature, 22
Zsigmondy's ultra-microscope, 227